中国传统服饰文化英译系列丛书

中华传统

京剧服饰文化翻译研究

张慧琴 著

Cultural Translation of Traditional
Peking Opera Costumes

中国纺织出版社

内 容 提 要

京剧作为中国文化的重要符号之一，2010年已被列入联合国《人类非物质文化遗产代表作名录》，在中华文化"走出去"的今天，其对外翻译传播的意义更加凸显。京剧服饰则以其靓丽的色彩、独特的款式、精美的图纹以及高超的工艺吸引着无数观众"走近"京剧，探索其深厚的文化意蕴。因此，京剧服饰文化翻译绝不是简单的文字转换，而是基于深厚文化积淀和高超语言功底的协调转换活动。

本书在回顾中华传统京剧国内外交流与研究状况的基础上，阐述了京剧服饰文化的对外交流，通过聚焦大衣箱、二衣箱和三衣类的文化内涵，引用国内外学者针对京剧服饰文化的全英或英汉双语表达，补充笔者的汉语"拙译"或汉英双语的"补译"与"拙译"，启发读者对照英汉平行文本，反思京剧服饰文化两种语言的不同表达，传承弘扬民族服饰文化，增强文化自信，助力文化"软实力"建设，推进文化"走出去"战略的实施。同时，基于原文与译文的对照剖析，尝试构建京剧服饰文化翻译协调理论，在"意、诚、度、心、神"的五步协调中，成就京剧服饰文化翻译协调之美。

图书在版编目（CIP）数据

中华传统京剧服饰文化翻译研究 / 张慧琴著 . — 北京：中国纺织出版社，2020.4

（中国传统服饰文化英译系列丛书）

ISBN 978-7-5180-5705-4

Ⅰ.①中⋯　Ⅱ.①张⋯　Ⅲ.①京剧—剧装—服饰文化—英语—翻译—研究—中国　Ⅳ.① TS941.735

中国版本图书馆 CIP 数据核字（2018）第 280688 号

策划编辑：李春奕　　责任编辑：谢冰雁
责任校对：寇晨晨　　责任印制：王艳丽

中国纺织出版社出版发行

地址：北京市朝阳区百子湾东里 A407 号楼　邮政编码：100124

销售电话：010 — 67004422　传真：010 — 87155801

http://www.c-textilep.com

中国纺织出版社天猫旗舰店

官方微博 http://weibo.com/2119887771

北京玺诚印务有限公司印刷　各地新华书店经销

2020 年 4 月第 1 版第 1 次印刷

开本：787×1092　1/16　印张：14.5

字数：274 千字　定价：69.80 元

序

preface

　　自1930年我国京剧表演艺术家梅兰芳先生远渡重洋，将最具代表性的国粹——中华京剧艺术带到了美国，从此京剧艺术开始在世界各地传播，其迷人魅力促进了中外文化的交流。

　　近年来，随着我国综合国力的增强，人民生活水平不断提高，人民对文艺作品在内的文化产品的质量、品味和风格等的要求也不断提升。中华文化瑰宝京剧艺术再度引起人们空前的关注。京剧服饰作为京剧艺术中博人眼球的靓丽载体，吸引了无数国内外京剧爱好者"走近"京剧，聚焦京剧服饰文化艺术。如何帮助熟谙京剧的"行家"用英语介绍京剧，助力京剧服饰文化"走出去"；同时，又如何帮助京剧爱好者在学习京剧服饰文化的过程中，从不同角度理解英汉语言文化相互的转换规律，逐步掌握京剧服饰的英汉双语表达，在换位思考中尊重、理解、协调文化差异，成就京剧服饰和谐美的译文，成为本书写作的缘起。

　　全书引用诸多中外学者针对京剧服饰的全英或英汉双语表达，补充笔者的汉语"拙译"或汉英双语的"补译"与"拙译"，比较其不同的表达方式，启发读者主动思考，不仅要知其然，更要知其所以然。虽然本书在案例剖析中并没有针对京剧服饰术语的英汉双语文本逐一对比剖析，但是所有章节中有关京剧服饰大衣箱、二衣箱和三衣类的英汉表达，无一不是遵循"意、诚、度、心、神"的五步协调，并最终统一到"真、善、美"京剧服饰文化翻译协调美的有力例证。正是基于上述原因，促成了京剧服饰文化翻译协调理论构建的探索。

　　文化如水，润物无声。京剧作为全人类共同的文化遗产，在传播中国文化精神的同时也促进世界人民心灵的沟通。祈愿本书的文化翻译探索能助推京剧服饰文化艺术跨越文化差异，使不同文化在"和而不同"中相互理解和相互尊重、在"求同存异"中相得益彰、在交流互鉴中增进友谊，并促进彼此的交流与发展。

该著为北京市哲学社会科学规划重点项目（15JDWYA008）以及北京市"长城学者"项目的部分研究成果，在此一并致谢！

张慧琴

2019年8月1日

于北京樱花园甲2号

目 录
c o n t e n t s

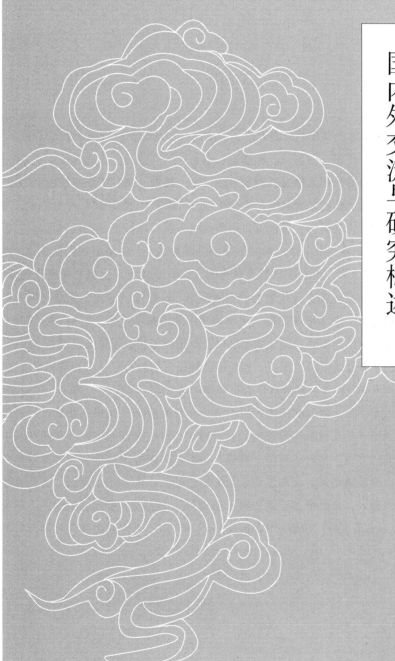

第一章

中华传统京剧
国内外交流与研究概述

第一节
引言

　　京剧是中华优秀文化的代表之一，历史悠久，从清代乾隆五十五年（1790），江南四大徽班——三庆班、四喜班、和春班、春台班的入京，到道光年间（1821—1850），宫廷戏与民间戏曲的相互借鉴，再到吸纳汉剧、昆曲、秦腔以及民间曲调与表演方法，逐步使京剧形成完整的艺术风格与表演手法，❶ 并在19世纪末至20世纪初，成为中国最具生命力和艺术表现力、最能体现民族风格且最具国际影响力的戏曲。2008年，教育部在全国10个省（市）中小学开展京剧进课堂试点，传承中华优秀文化，增强学生文化自信；2010年，京剧成功列入联合国《人类非物质文化遗产代表作名录》。与此同时，京剧中颇具民族文化代表性的京剧服饰，在塑造戏曲角色形象与展现角色身份地位、时代背景、生活地域环境、民族风情、个性特征及生活习俗等的同时，也以其靓丽的色彩、独特的款式、精美的图纹以及高超的工艺吸引着无数观众"走近"京剧，成为京剧表演中最为吸睛的组成。如何借助翻译，跨越时空，实现京剧服饰文化交流，对于弘扬民族文化、增强民族自信以及实施中华文化"走出去"战略，意义重大。

　　基于上述原因，本书在国内外京剧文化和京剧服饰文化史的基础上，针对京剧服饰文化术语给予英汉两种不同语言的归类对比，探究京剧服饰术语文化艺术中英汉语言的不同表达方式。在一定程度上遵循外语教学与研究出版社博雅教育双语系列出版物的宗旨，满足英语学习者和汉语学习者通过阅读中国主题（京剧）读物提高英语和汉语能力的需求，又以中英双语思维、架构和写作的形式予以后世学人以启迪。维特根斯坦（Ludwig Josef Johann Wittgenstein）有云："语言的边界，乃是世界的边界"，诚哉斯言。

---

❶ 赵晴. 借助戏曲艺术传播传统文化［J］. 文学前沿，2008(2)：111–122.

第二节
中华传统京剧对外文化交流概述

回顾历史，中华传统京剧艺术作为中外文化交流的重要载体，其对外传播主要经历了三个阶段。

第一阶段：从光绪十七年（1891）开始到20世纪40年代末。光绪十七年，京剧武生演员张桂轩就曾带领戏班到日本长崎、大阪、东京和神户等地演出，后在光绪二十一年（1895）和光绪二十二年（1896）先后到朝鲜的汉城（今韩国首尔）和俄罗斯的海参崴（今俄罗斯符拉迪沃斯托克）等地演出。清末民初，杨月楼、黄月山、王鸿寿、绿牡丹、唐韵笙、高百岁、李多奎、高庆奎、周信芳、王宝魁、王洪涛、王少奎、周喜增、张桂轩、刘竹友、碧玉蝉等京剧艺人都曾到俄罗斯海参崴等地演出，甚至常年在海参崴的颂竹舞台和伯力（今俄罗斯哈巴罗夫斯克）的伯力大戏院演出京剧。与此同时，京剧在东南亚的演出也很频繁。以新加坡为例，宣统二年（1910），福州新和祥班八十多人在新加坡庆升平戏院演出京戏《水帘洞》《金刀阵》《金钱豹》《四杰村》和《花蝴蝶》等。后来随着庆升平戏院改为庆升平新舞台，新和祥班又在新加坡演出了《铁公鸡》《路遥知马力》《连环套》《落马湖》等剧目。1931年（民国二十年）前后，新加坡新世界游艺场中的大舞台一直以京剧为主要上演剧目，如《目莲救母》《少林寺》《火烧汉阳城》《枪毙阎瑞生》《天雨花》《临江驿》《玉堂春》和全本《粉妆楼》等常年连续上演。甚至居住在北京和上海等地的花旦白玉艳、老生白叔安、花旦徐碧霞、武生盖俊廷等人都曾在新加坡演出过《荒江女侠》《女侠红蝴蝶》《白蛇传》《大英杰烈》《狸猫换太子》等剧目。抗日战争前后，京剧演出活动在新加坡也十分繁盛。

在众多京剧的海外传播中，梅兰芳的访美演出堪称最具影响力。早在1919年访日期间，梅先生就曾演出《天女散花》《御碑亭》《黛玉葬花》《虹霓关》《贵妃醉酒》《尼姑思凡》《琴挑》《游龙戏凤》《武家坡》《嫦娥奔月》《春香闹学》和《游园惊梦》等剧目，并在1924年10月再度访日时演出了系列京剧，包括《贵妃醉酒》《洛神》《廉锦枫》《红线盗盒》《御碑亭》和头本《虹霓关》等梅派代表性剧目，极大地丰富了京剧的表演手段，充分体现出中国戏剧美学的特色和文化内涵。此后黄玉麟和小杨月楼，还包括郑法祥、张玉峰、董志扬、筱九霄、马金凤、蒋月楼等艺人，都曾于1925年（民国十四年）和1926年（民国十五年）期间赴日演出，扩大了京剧在日本的影响，增进了中日两国戏剧文化的交流与融合。1930年，梅兰芳与齐如山等人乘加拿大皇后号由上海前往美国各地，在纽约演出《汾河湾》《剑舞》《刺虎》等剧目，随后在西雅图、芝加哥、旧金山、洛杉矶、圣地亚哥和檀香山等地演出了《芦花荡》《青石山》《打隍城》《空城计》《汾河湾》《贵妃醉酒》《打渔杀家》《春香闹学》《刺虎》《虹霓关》

3

《廉锦枫》《天女散花》《霸王别姬》《红线盗盒》《西施》《麻姑献寿》《嫦娥奔月》《上元夫人》等京剧72场，使中国京剧文化首次得到西方的认同，同时也使京剧艺术开始成为世界性的艺术。1935年梅兰芳访美归来，应苏联对外文化协会的邀请，演出京剧《汾河湾》《刺虎》《打渔杀家》《宇宙锋》《虹霓关》《贵妃醉酒》，以及《红线盗盒》中的剑舞、《西施》中的羽舞、《麻姑献寿》中的袖舞、《木兰从军》中的戟舞、《思凡》中的拂尘舞、《抗金兵》中的戎装舞；还有《青石山》《盗丹》《盗仙草》《夜奔》和《嫁妹》等京剧片断，引起了苏联戏剧家斯坦尼斯拉夫斯基（Konstantin·Stanislavsky）、德国戏剧界泰斗贝尔托·布莱希特（Bertolt Brecht）和著名文学家玛克西姆·高尔基（Maxim Gorky）、阿·托尔斯泰（Alexei Nikolayevich Tolstoy）等对中国传统戏曲和以梅兰芳为代表的京剧表演体系的关注，在一定程度上消除了当时西方对中国戏剧甚至中国人的偏见，促进了中国京剧在海外的传播，增进了东西方文化的交流，使京剧艺术这一中国文化的奇葩在世界戏剧舞台上大放光彩。

第二阶段：从20世纪50年代到70年代末。新中国成立以后，我国高度重视京剧艺术的文化交流与传播，多次派遣京剧团赴世界各地访问演出。梅兰芳先后于1952年、1957年和1960年赴苏联访问，并在1956年第三次访问日本，为中国京剧传统文化在海外赢得了巨大荣誉。1951年，张云溪、张春华等7人参加在德国柏林举办的第三届世界青年与学生和平友谊联欢节，演出京剧《三岔口》《武松打虎》《水帘洞》和《红桃山》四出武戏。这是新中国成立后京剧艺术首次以国家名义在国外演出。1953~1962年期间，中国组成的京剧演出团又参加了第四届~第八届世界青年与学生和平友谊联欢节，先后演出了《雁荡山》《秋江》《春郊试马》《虹桥赠珠》《游园惊梦》《水漫金山》《拾玉镯》和《断桥》等剧目。此外，在世界青年联欢节上演出的《猎虎记》《挡马》《小放牛》《盗仙草》《猪婆龙》《打焦赞》《拾玉镯》《嫦娥奔月》《哪吒闹海》等京剧剧目倍受欢迎。后又因国家领导人出访或剧团应邀，先后在美国、联邦德国、比利时、英国、卢森堡、荷兰、希腊、土耳其、突尼斯、朝鲜、日本和加拿大等多个国家演出了《杨门女将》《三岔口》《雁荡山》《火凤凰》《秋江》《贵妃醉酒》《除三害》《拾玉镯》《挡马》《白蛇传》《闹天宫》《昭君出塞》《水漫金山》《探谷》《岳母刺字》和《野猪林》等剧目，这些演出使京剧的海外影响力迅速扩大。同时，京剧演员也常随外交使团出访到海外进行演出，足迹遍及了印度、缅甸、委内瑞拉、哥伦比亚、古巴、加拿大等国家。中国京剧在海外的影响力与日俱增，海外观众也乐于通过欣赏京剧来认识中国文化、认识中国的戏剧艺术，京剧成为我国文化外交的重要手段。

第三阶段：从1978年改革开放至今。随着我国经济的发展以及对外开放政策的实施，文化交流从形式到内容不断创新，京剧艺术在海外的影响力也逐渐扩大。中国国家京剧院、北京京剧院、上海京剧院、天津青年京剧团等著名京剧团体和一些地方京剧院团也相继加入海外传播的行列，从不同角度拓展了京剧的发展空间与接受渠道。

2010年，京剧被联合国教科文组织列入《人类非物质文化遗产代表作名录》，并且作为"国剧海外传播工程"和"百部国剧英译工程"的国家级项目，开始多渠道走出国门，走遍世界各地，成为介绍和传播中华文化的重要途径。

时值当下，新媒体、大数据的出现给京剧文化的传播带来了革命性转变，互联网传播改变了我们对于京剧传播原有的思想观念、决策指导及行动方式。以京剧网站——中国京剧网（http://www.jingju.cn）的开发和建设为例，网站的界面设计以京剧服饰文化相关的视觉元素为主，内容包括京剧服饰种类、样式、图案、颜色、头饰和脸谱等相关图片和图案，以及京剧服装知识，包括对京剧服饰文化元素的挖掘和归纳等。这些都使京剧的传播速度、范围和信息量快速实现了跨地域、跨民族的传播，其影响范围更广、受众群体更多、更具感染力，且更易被大众群体理解和接受。同时，中国互联网络信息中心（CNNIC）于2018年12月发布的第43次《中国互联网络发展状况统计报告》（以下简称为《报告》）称，截至2018年12月，中国网民规模达到8.29亿，互联网普及率为59.6%。可见网络已经成为文化传播的新途径，网络传播的广泛性也使得京剧文化的传播打破了地域限制而具有了全球性。网络信息的海量性可以丰富京剧文化数字信息库的建设；网络信息的实时性可以快捷地把京剧文化的相关新闻及时推送到公众手中；同时，网络的声画结合可以通过图片、文字、视频等形式立体地展示京剧内容以及京剧脸谱、服饰等，在提升公众兴趣的同时，更能借助数字媒体来保护和传承京剧文化内涵。

无独有偶，笔者在查阅国内对外宣传中有关京剧的网站时，通过摘录"首都之窗——北京市人民政府"英文网站和"Top China Travel"英文网站中关于京剧的介绍，梳理了其相关内容，拟从英汉对照表达中反思传统京剧艺术的对外传播现状（英文表达之后的中文部分为笔者拙译）。

原文　Beijing Opera[1], also called "Eastern Opera", is a principle tradition in Chinese culture. It is called Beijing Opera because it is formed in Beijing.

Beijing Opera has a history of 200 years in which its fountainhead can be dated back to old local operas, especially Anhui Opera, which was very popular in northern China in the 18th century. In 1790, the first Anhui Opera performance was held in Beijing to celebrate the Emperor's birthday. Later, some other Anhui Opera troupes went on to perform in Beijing. Anhui Opera was easy to move and good at absorbing the acting styles of other types of operas. Beijing accumulated many local operas, which made Anhui Opera improve quickly.

At the end of the 19th century and the beginning of the 20th century, after merging for 10 years, Beijing Opera finally formed, and became the biggest of all operas in China. Beijing Opera

---

[1] 本书中有关"京剧"一词的英译，有几种形态：Beijing Opera, Peking Opera, Chinese Opera, Jingju等，为尊重原著保留原文。

has a rich list of plays, artists, troupes, audiences, and wide influences, making it the foremost opera in China.❶

拙译 京剧是中华传统文化之本，又被称为是"东方歌剧"，取名"京剧"是因为其形成于北京。

京剧约有二百年的历史，其缘起可追溯到古老的地方戏曲，特别是18世纪时在中国北方广受欢迎的徽剧。1790年，在北京演出的第一部徽剧原本只是为皇帝祝寿。后来，因徽剧曲调多变，又具有擅长吸纳其他剧种的动作风格等特点，一些戏班在北京得以继续演出。同时，北京当时汇集了许多不同的地方戏曲，这都使徽剧可以取诸家之长而快速提升。

19世纪末20世纪初，历经十余年的发展，京剧最终得以形成。作为我国最大的剧种之一，京剧剧目丰富，产生了大批优秀的表演艺术家和戏班子，并吸引了无数的京剧迷，影响广泛，成为全国位居首位的剧种。

原文 Beijing Opera is a comprehensive acting art. It blends singing, reading, acting, fighting, and dancing together by using acting methods to narrate stories and depict characters. The roles in Beijing Opera include the male, female, painted-face, and comedic roles. Besides, there are other supporting roles as well. In addition, the types of facial make-up, especially concerning the color, are the most particular art in Beijing Opera, because they can symbolize the personalities, characteristics and fates of the roles.❶

拙译 京剧属于综合的表演艺术，集唱、念、做、打、舞于一体。唱——包括演员的独唱和对唱。大段独唱叙述人物的背景和对命运的慨叹，表现人物复杂、矛盾的思想感情，从而深入刻画人物的内心世界和性格。对唱往往起到推动情节进展、将矛盾冲突引向高潮的作用。❷念——指演员在舞台上的念白。念白又分为两种：一种是在语言的节奏、韵律和腔调上经过加工的、以中州韵为基础的"韵白"，它与日常语言有相当的距离，并已有一定程度的音乐化。"韵白"常表现上层人物的身份和其具有的庄重性格。另一种念白是以各剧种所使用的方言为基础的"散白"（道白），与日常语言比较接近，多用于花旦和丑角，主要表现下层人物的身份及其活泼轻佻的性格。不论是"韵白"还是"散白"，都是经过艺术提炼的语言，都要求演员通过念白恰当而真实地表现出人物的性格、情感和身份。戏曲界流行的"千斤话白四两唱"的说法，表明了念白在戏曲表演中的重要作用。做——一般是指舞蹈化的形体动作。演员在角色塑造时，手、眼、身、步各有程式；髯口、翎子、甩发、水袖有多种技法，同时在运用中还要注意表现出特定角色的性格、年龄、身份等特征。打——指戏曲对传统武术

---

❶ China-Fun. The History of Beijing Opera［N/OL］. eBeijing, 2010-11-04. http://www.ebeijing.gov.cn/BeijingInformation/BeijingsHistory/t1137406.htm

❷ 王慧. 戏曲演唱的若干要点［J］. 戏剧之家，2014(16)：51.

的舞蹈化，一般分为"把子功"和"毯子功"两类。凡用古代刀、枪、剑、戟等兵器（习称"刀枪把子"）对打或独舞的，称为"把子功"；在毯子上做翻、滚、跌、扑等动作的，称为"毯子功"。戏曲表演中"唱、念、做、打"有机结合，融为一体，共同为表现剧情和塑造角色服务。❶

原文 It is widely acknowledged that the end of the 18th century was the most flourishing period in the development of Beijing Opera. During this time, there were lots of performances not only in folk places, but also in the palace. The noble class loved Beijing Opera; the superior elements in the palace played a positive role in the performances, make-up, and stage setting. The mutual influence between palace and non-government places promoted Beijing Opera's development.

From the 1920's to the 1940's of last century was the second flourishing period of Beijing Opera. The symbol of this period was the emergence of lots of sects of the opera. The four most famous were "Mei" (Mei Lanfang 1894−1961), "Shang" (Shang Xiaoyun 1900−1976), "Cheng" (Cheng Yanqiu 1904−1958), and "Xun" (Xun Huisheng 1900−1968). Every sect had its groups of actors and actresses. Furthermore, they were extremely active on the stage in Beijing, Shanghai and so on. The art of Beijing Opera was very popular at that time.❷

拙译 众所周知，18世纪末是京剧发展最为繁荣的时期。这一时期，除了在民间，包括宫廷里也有很多京剧表演。贵族阶层对于京剧的喜爱使京剧在戏曲表演、化妆、舞台设置中都更受关注。宫廷与非官方的相互影响共同促进了京剧的发展。

20世纪20~40年代是京剧发展的第二个繁荣时期。这一时期，出现了许多具有代表性的派别。最著名的有"梅"（梅兰芳，1894—1961）、"尚"（尚小云，1900—1976）、"程"（程砚秋，1904—1958）和"荀"（荀慧生，1900—1968）这四派。每个派别都有其特定的演员团队，而且他们在北京、上海等地的舞台演出非常活跃。毫无疑问，京剧艺术在当时很受欢迎。

原文 Born in Beijing in 1894 into a family full of Chinese Opera performers, Mei Lanfang was the best known Chinese Beijing Opera master. In his 50-year stage career, his exclusive skills of females characters playing has won him a world wide acclaim. He was also one of the "Four Great Dan", along with Shang Xiaoyun, Cheng Yanqiu and Xun Huisheng in Beijing Opera history. Mei Lanfang was the first Beijing Opera master who participated in cultural exchanges with foreign countries in his contemporary time in China. During 1919 to 1935, he paid visit to Japan,

❶ 张亚秋. 戏曲表演的基本功——唱念做打 [J]. 现代交际, 2009(5)：92-93.
❷ China-Fun. The History of Beijing Opera [N/OL]. eBeijing, 2010−11−04. http://www.ebeijing.gov.cn/BeijingInformation/BeijingsHistory/t1137406.htm

中华传统京剧国内外交流与研究概述 第一章

7

United States and other regions, spread Beijing Opera to foreign countries. Forming friendships with western artists, showing the harmony and beauty in Chinese performance had played a important role to eliminate the bias between China and West in that era.❶

拙译 我国最著名的京剧艺术家——梅兰芳, 1894年生于京剧世家。在从事舞台艺术50年之际, 他对女性角色的独特塑造使他赢得了全世界的赞誉, 并且同尚小云、程砚秋、荀慧生三位艺术家一起成为我国京剧史上的"四大名旦"。作为当时第一位与外国进行文化交流的京剧艺术大师, 梅兰芳在1919~1935年访问了日本、美国和其他国家地区, 将京剧传播到了国外。其和谐优美的京剧表演, 以及与西方艺术家建立的友谊, 对于消除当时中国和西方彼此之间存在的偏见发挥了重要作用。

原文 As the story goes, Emperor Qianlong of the Qing Dynasty fell interested in the local drama during his inspection of the Southern China in disguise. To celebrate his 80th birthday in 1790, he summoned opera troupes from different areas around China to perform for him in Beijing. After the celebration, four famous troupes from Anhui Provinces were asked to stay, for audiences were particularly satisfied with their beautiful melodies, colorful costumes and interesting facial patterns.❶

拙译 有故事流传, 清代乾隆微服私访南方, 对京剧产生了浓厚兴趣。1790年为庆祝他的80岁大寿, 他命各地戏班入京表演。完成庆典后, 安徽的四大戏班凭借其悦耳的曲调、多彩的戏服和有趣的脸谱, 得以继续留京演唱。

原文 Gradually it replaced Kunqu Opera which had been popular in the palace and among the upper ranks in Beijing. Later, some troupes from Hubei Province came to Beijing and often performed together with the Anhui troupes. The two types of singing blended on the same stage and gradually gave birth to a new genre that was known as Beijing Opera.❶

拙译 在皇宫的高级官员中, 徽剧倍受欢迎, 逐渐取代了昆曲的位置。后来, 湖北的戏班来到北京, 并经常和安徽的戏班一起演出, 这两种唱风相互融合, 逐渐促成了新剧种——京剧的诞生。

原文 Beijing Opera absorbed various elements of its forerunners, such as singing, dancing, mimicry and acrobatics, and adapted itself in language and style of singing to Beijing audiences. As time goes by, its popularity spread all over the country, becoming the most popular drama in China.Furthermore, Beijing Opera sing has influenced Chinese drama deeply.❶

拙译 京剧吸纳了很多前卫的元素, 诸如唱歌、跳舞、口技和杂耍, 并把这些元

---

❶ Top China Travel. Beijing Opera [N/OL]. http://www.topchinatravel.com/china-guide/peking-opera.htm

素运用于其语言和演唱风格之中。后来，京剧逐渐在全国推广，并成为中国最受欢迎的剧种。而且，京剧的唱腔也对我国戏曲产生了深远影响。

原文 Elaborate and gorgeous facial make-up and costumes are two distinguished characteristics of Beijing Opera. The audience can know what kind of character the role is from the colors and patterns.❶

拙译 精致而耀眼的脸谱和戏装是京剧的两大特色。借助不同的色彩和脸谱造型，观众可以了解戏曲人物的角色。

原文 Generally speaking, red faces have the positive meaning symbolizing the brave, loyalty, upright and wise men. Another positive color is purple. Black faces usually have neutral meaning, representative the just men and uprightness. Blue and green also have neutral meanings that symbolize the hero from the bushes or some kind of rebelliousness leaders. Meanwhile, the yellow and white represent the crafty men with negative meaning such as treachery and fierce hearts. Performers have gold or silver facial make-up standing for the monsters or Gods and super natural power. Good-nature people are usually painted with relatively simple colors while make-up of hostile and doubtful characters, such as bandits, robbers, rebels and alike, bear complex marks.❶

拙译 一般而言，红脸是正面人物，象征勇敢、忠诚、刚正和智慧；黑脸通常是中立的意思，象征正直和诚实。还有一个正面角色的代表色是紫色。蓝色和绿色也是中立颜色，象征着草莽英雄或反叛领袖。此外，黄色和白色象征着奸诈，是反面人物的象征，如叛徒和黑心肠的人。如果脸谱是金色或银色则象征着超自然力的怪物或神仙。好心肠的人通常都是相对简单的脸谱色，而那些充满敌意和可疑的人物，如强盗、盗贼、反叛者等，都有复杂的标记。

原文 The costuming of Beijing is based mainly on the court and civil costumes of the Ming dynasty style, with frequent uses of deep red, green, yellow, white black and blue. Strong contrasting colors are freely used and embroidered in gold, silver and colored threads. The rules for costumes are strictly based on rank, occupation and life style. And there are special costume with different colors and designs for each role.❶

拙译 京剧的服装主要基于清代宫廷和民间的服饰风格，频繁使用深红色、绿色、黄色、黑白色和蓝色，并对强烈的对比色运用自如，同时配有金色、银色和彩色刺绣。其规矩严格遵守官衔等级、职业和生活模式，每个角色都有特制的戏服和不同颜色的脸谱对应。

---

❶ Top China Travel．Beijing Opera［N/OL］．http://www.topchinatravel.com/china-guide/peking-opera.htm

第三节
中华传统京剧国内外研究概述

关于京剧艺术的对外交流正如上一节所述，历经三个阶段的发展，从传统模式到现代互联网的新媒体技术，包括对外宣传英文网站的建设等，都使京剧文化艺术在海外的受众面和影响力得到前所未有的扩展。同时，中华传统京剧艺术在海外传播历经数百年，已经引起西方学者的关注。基于美国哈佛大学学者卞赵如兰（Rulan Chao Pian）《京剧研究在西方》的阐述，笔者聚焦20世纪80年代中期以前的欧美（日本、俄国和东欧除外）对于京剧的研究状况，综合前人研究成果，将西方学者对于中华传统京剧的研究大致归纳为以下六个方面。

一是选择从我国历史上元、明、清三个朝代不同的时间段为切入点，对京剧给以整体性和概括性研究。以20世纪20年代的苏烈模、祖克和巴斯等汉学家为学界代表，针对戏剧术语，给以逐条解释。二是关注京剧历史沿革与近代发展，主要以马克林（Colin Patrick Mackerras）为代表的学者，结合我国戏剧历史，分别以历史上元、明、清三个朝代和近代不同的时间点为坐标，尝试从社会学的角度研究京剧在不同阶段的发展。三是记录京剧演员的生活，诸如英国戏剧学者斯科特（Adolphe Clarence Scott）的《梅兰芳：中国梨园界泰斗》和《舞台生活四十年》等，帮助读者通过对艺术家生活的了解，理解京剧艺术。四是基于西方读者对京剧创作的艺术性理解，研究我国京剧的创作方向。五是探索京剧剧本的翻译研究，从早期故事梗概的翻译，到20世纪60年代整个剧本的全译，旨在借助文字把舞台上的动作、音乐效果给以最大限度的表达。其中针对京剧翻译研究颇具代表性的学者有卞赵如兰，其主要贡献是基于录音翻译了三出戏（《打渔杀家》《捉放曹》《苏三起解》）；美国夏威夷大学教授魏丽莎（Elezabeth Ann Wichmann）则提出京剧翻译的基本方法，倡导对原作要从内涵和精神到风格都原汁原味地保留，甚至对原有的形象化比喻都依原文顺序给予保留；黄为淑（Hwang Wei-shu）在其博士论文中结合了诸多的翻译案例，对戏剧翻译理论、翻译原则以及翻译程序都进行了详细阐述，强调以演出为目的的翻译策略，特别是剧本唱词、对白、科介说明、上下场诗、双关语以及专有名词等各个方面。同时，黄博士还将京剧翻译过程概括为：基于自己的兴趣与学术能力选好合适的剧本；查阅并掌握剧本的相关背景知识，比较剧本的所有现译本；完成详尽的自译初稿；基于舞台演出和出版为目的，在大声朗读中感受译本的风格与节奏；反复审读、修正译文初稿。六是聚焦京剧曲谱或音乐专题的研究。颇具代表性的是美国魏丽莎博士的《倾听戏剧：京剧的听觉维度》，以及民主德国（1949—1990）的沈费德有关京剧唱腔结构的论著。

无独有偶，中国学者对于中华传统京剧的研究内容则基本从历史渊源❶到发展轨迹、从表演艺术到对外传播、从学校课程构建❷❸到学生综合素质培养❹、从中国京剧与西洋歌剧或日本歌舞伎的比较❺❻❼到中国京剧在海外的传播策略剖析❽❾。

基于上述西方学者对于中华传统京剧六个不同方面的研究以及国内学者对于京剧文化的多角度聚焦，笔者在概述中将选取塑造角色外部形象和展示角色身份、年龄、个性及生活习俗的服饰——京剧中的"行头"作为研究对象；并在诸多有关京剧服饰文化翻译的学者中，选取谭元杰编著的《中国京剧服装图谱》（立足传统京剧人物造型整体形象，用线描的形式系统展现传统京剧服装的基本形式及主要服装品种）、赵少华编著的《中国京剧服饰》（选取京剧服饰中最具有代表性的作品，简要介绍京剧服饰的不同类别、使用途径及其艺术文化特色）、阚艳华等编著的《中英对照京剧服饰术语》和美国人亚历山德拉·B.邦兹（Alexandra B.Bonds）编写的 Beijing Opera Costumes: The Visual Communication of Character and Culture（聚焦20世纪末至21世纪初京剧艺术，关注并阐释其传统服饰文化），以及潘霞凤在新世界出版社出版的 The Stagecraft of Peking Opera（采用英汉双语对照的表述方式聚焦京剧服饰，依照京剧传统服装的分类标准。除蟒、帔、靠、褶四大类之外，其余的所有剧装被统称为衣。而衣的品种若按照服装本身形制的基本特征又可分为长衣、短衣、专用衣和配件四部分；若按照衣箱的管理制度归纳分类则可分为大衣箱、二衣箱和三衣箱）。限于篇幅，本书拟针对大衣、二衣和三衣的文化内涵，结合京剧服饰文化协调美的翻译原则，给以多角度阐释，并在探索与构建京剧服饰文化翻译理论的过程中，论证和实践该理论的可行性与科学性。

本章小结 ｜ 京剧作为中华民族一个古老而传统的剧种，历史悠久。近年来，随着中华文化"走出去"战略的实施，文化"软实力"的建设倍受重视。京剧作为中华民族精神文化的重要组成，其服饰作为重要的物质文化载体也逐渐走入研究者的视野，成为文化艺术传承与传播的有效途径之一。本章概述了中华传统京剧国内外交流与研究，并采用英汉两种不同语言的表达方式介绍了京剧的历史、特点与发展。同时，梳理了国内外学者对于中华传统京剧的研究现状，并阐述了本书的研究内容、方法与目标。

❶ 王钟陵. 中国京剧史略论［J］. 清华大学学报（哲学社会科学版），2008(3)：5–17.
❷ 黄敏. "小京迷"特色乡村课程探析［J］. 上海教育科研，2018(6)：84–88.
❸ 江峰，徐家瑞. 中小学京剧课程建设的范式探索——培养学生艺术素养的视角［J］. 中国民族教育，2018(6)：43–45.
❹ 胡淳艳. 高校京剧教材及教育状况杂谈［J］. 中国大学教学，2010(10)：83–84.
❺ 苏冬花. 中国京剧与日本歌舞伎的比较研究［J］. 戏曲艺术，2005(2)：20–24.
❻ 陈晓芸. 中国京剧与西洋歌剧之比较［J］. 大舞台，2012(9)：1–2.
❼ 余意梦婷. 中国京剧与西方歌剧之审美差异——以中国京剧《图兰朵公主》和西方歌剧《图兰朵》为例［J］. 戏剧文学，2014(10)：28–32.
❽ 于晓华. 探究构建京剧艺术对外传播的科学模式［J］. 戏剧研讨，2018(8)：51.
❾ 凌来芳. 中国戏曲"走出去"译介模式探析——以"百部中国京剧经典剧目外译工程"丛书译介为例［J］. 戏剧文学，2017(8)：117–124.

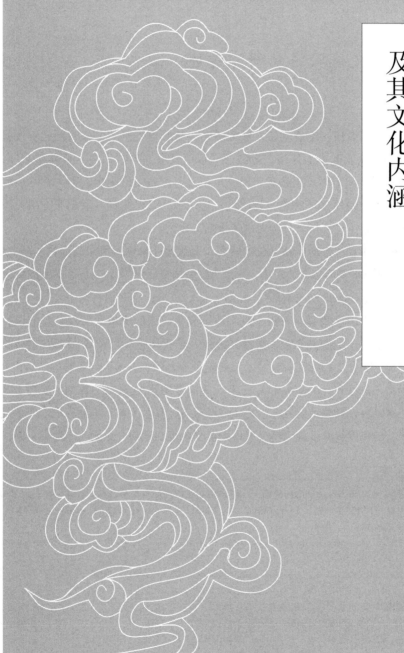

第二章

京剧服饰文化传播及其文化内涵

第一节
京剧服饰文化国内外研究综述

京剧形成于清代，随着剧情发展的需要，其服饰基于唐、宋时期代表性服饰的特点，同时吸纳满族服饰的风格，最终形成以明代服饰款式为主的京剧服饰——行头，同一套行头在不同的剧目里表现不同的角色，可以映射出剧中人物的性格。京剧行头通常可以在不同的情形下穿着，无时代、地域或季节之限制。京剧服饰研究对于理解京剧剧情、加强传统京剧文化的对外传播意义重大。

纵观国内外有关京剧服饰文化的研究、聚焦国外学者对于京剧服饰的研究后发现，目前的京剧服饰研究可以分为三类：一是京剧服饰史研究❶❷，二是京剧服饰文化研究❸，三是京剧服饰材质与形制研究❹❺。

同时，笔者梳理我国学者对于京剧服饰文化研究的成果后发现，目前研究大致也可以分为三类：一是图文并茂，中英文结合，详细介绍京剧服饰中的五大类戏衣（蟒、帔、靠、褶、衣）及其穿戴规制❻❼；二是选取京剧服饰中最具代表性的作品，介绍京剧服饰的种类，包括大衣（袍服类）、二衣（武装铠甲类）、三衣（内衣、鞋靴）及盔帽的用途与艺术特色，聚焦京剧衣箱、行头与穿戴法，甚至妆容，描述京剧衣箱的发展演变过程❽❾❿；三是以经典京剧为例，采用中英文对照，详细介绍剧目、曲谱与曲词⓫⓬。

---

❶ DOLBT W. A History of Chinese Drama［M］. London：Paul Elek，1976.

❷ ARLINGTON L C. The Chinese Drama：from the Earliest Times until Today［M］. Shanghai：Kelly and Walsh，1930.

❸ BONDS ALEXANDRA B. Beijing Opera Costumes：the Visual Communication of Character and Culture［M］. Honolulu：University of Hawaii Press，2008.

❹ JACOBSEN R D. Imperial Silks：Ch'ing Dynasty Textiles in the Minneapolis Institute of Arts［M］. Minneapolis：Minneapolis Institute of Arts 2000.

❺ CAMMANN SCHUYLER. China's Dragon Robes［M］. New York：Ronald Press Co，1952.

❻ 谭元杰. 中国京剧服装图谱［M］. 北京：北京工艺美术出版社，2008.

❼ 张大夏. 国剧行头［M］. 台湾：国立台湾艺术馆，1961.

❽ 刘月美. 中国京剧衣箱［M］. 上海：上海辞书出版社，2002.

❾ 赵少华. 中国京剧服饰［M］. 北京：五洲传播出版社，2004.

❿ 刘琦. 京剧行头［M］. 天津：百花文艺出版社，2008.

⓫ 孙萍. 中国京剧百部经典英译系列［M］. 北京：中国人民大学出版社，2012.

⓬ 阙艳华，董新颖，王娜，等. 中英文对照京剧服饰术语［M］. 覃爱东，陶西雷，校译. 北京：学苑出版社，2017.

综上所述，目前国内外京剧服饰文化研究大多关注历史发展、大类服饰、剧本个案和特色人物；中英文对比版本的研究则集中在剧目的概要介绍、曲谱与词曲的探究，间或也有术语的梳理，但是有关京剧发展和衣箱分类，特别是关于文化内涵阐释的不同版本的中英文对照剖析相对鲜见。本书拟从京剧的发展入手，阐释京剧衣箱分类，关注其文化内涵，针对京剧传播现有的多种中英文版本进行归类梳理，逐一对照比较，尝试借助京剧服饰文化的协调翻译原则，补充完善当今文化语境下的传统京剧服饰文化内涵，增强文化"软实力"建设，促进京剧的传播与发展。

第二节
京剧服饰文化海外传播剖析

京剧形成于清代，其服饰包括唐宋时期一些具有代表性的服饰，但主要还是明代的服饰款式。随着京剧剧情发展的需要，其服饰种类不断丰富，并吸纳了满族的一些服饰特色，在繁缛刺绣与金属搭配的部件、色彩缤纷的脸谱和手工艺术等方面都展示出了服饰造型之美，甚至每套服装都映射出剧中人物的性格。早年在海外流传的有关京剧服饰的英文介绍对此都有论述，笔者尝试采用英汉对照文本的方式以飨读者，具体如下。

原文 Peking Opera compensates for the simplicity of the stage and props with very complex, and usually highly colorful and splendid, costumes and headgear. As with so much else in the Peking Opera, each costume is designed to fit the gender, personality, calling and status of character. Other mismatching features are of little importance: all costumes are based on the clothing of the Ming dynasty, no matter in which period the play is set, and characters wear the same costumes no matter the season in which the story takes place.[1]

拙译 京剧借助形式多样、多姿多彩的服饰和头饰，弥补其舞台和道具的简洁。如同诸多其他物品一样，每件服饰的设计都是为了适应剧中不同人物的性别、性格、职业和社会地位。所有服饰都是以明朝服饰款式为基础，其穿着不受地域或季节的限制，无论哪个时期，也不管故事发生在哪个季节，剧中人物都穿着相同的服饰。

原文 One reason why the costumes are so magnificent is because of the prevalence of

[1] MACKERRAS C P. Chinese Theater: from Its Origins to the Present Day [M]. Honolulu: University of Hawaii Press, 1983: 45.

robes, especially ceremonial ones. Use of ceremonial robes was originally reserved for members of the royal family and the aristocracy, but the eighteenth-century emperor Qianglong, having seen them in use in a local Kunqu performance, ordered their use in drama. After a period the costumes of the Peking Opera became more and more elaborate, until in the 1930s the Shanghai actor Zhou Xinfang felt it necessary to simply and reform the men's robes.❶

拙译　京剧戏服如此重要，其原因是长袍，特别是礼服的盛行。礼服原本只为皇室、贵族和上层社会人士设计，但在18世纪时，乾隆皇帝看到长袍在当地昆曲表演中应用之后，就下令在京剧表演中也如法炮制。经过一段时间后，京剧服饰日渐精致。这种趋势一直持续到20世纪30年代，由上海演员周信芳提出简化革新男式长袍的必要性后才告一段落。

原文　Zhou's contribution is notwithstanding, the robes of the Peking Opera remain one of its most colorful and grandest features. The most magnificent of the ceremonial robes are those worn by an emperor, which are yellow with a large coiled dragon symbolizing imperial power. Ceremonial robes used as costumes include a circular belt, termed a "jade belt" (Yudai, a circular belt worn with ceremonial robes) in Chinese, always too large a circle to hold any clothes in place but symbolic of crowns with tassels hanging down.❶

拙译　周信芳的重大贡献在于：保留了京剧戏服多彩与庄重并举之特色。其中最壮观的是皇帝出席庆典时穿着的龙袍，黄色龙袍上盘绕着象征帝王权力的龙，腰间有特大的环形玉带（玉带是庆典礼服的环形带）装饰，其作用并非固定任何服饰，只是为了与皇冠的流苏匹配。

原文　Armour came into vogue in Peking Opera in the1960s, with the growth in significance of the military plays at that time. Like the robes, the armour of generals or other seniors officers in military plays is very elaborate and colourful, whether worn by male or female characters.Many generals, male or female alike, have four flags attached to their back to increase the impression of power and valour. There is also an astonishing variety in the shape, color and design of helmets, which add to the variety and fascination of the costuming.❶

拙译　20世纪60年代，随着战争题材戏剧的增多，盔甲在京剧院开始流行。像长袍一样，在战争类戏剧中，无论男女，将士或高级军官都穿戴着色彩艳丽的精美盔甲。许多将士，不论男女，都在背后插着四面小旗以增加气势和感染力。盔甲在造型、色彩，以及头盔的设计方面也千变万化，令人瞠目，这些都使戏服款式多样，魅力无穷。

❶ MACKERRAS C P. Chinese Theater：from Its Origins to the Present Day［M］. Honolulu：University of Hawaii Press, 1983：46.

原文　Though less elaborate than the robes or armour, the coats worn by a variety of characters ranging from literati to young ladies or poor women also illustrates the complexity of Peking Opera costumes. Although these come in many colors, the bluish coats of poor women are of special interest, in that term "bluish clothes" (Qinyi, this character walk by way of vary quick but very short steps which, when performed properly, give the impression of grace and femininity) has come to be used for the demure, although not necessarily poor, female characters. For some female roles the "flowing sleeve" (Shuixiu) attached to the coat is highly significant. A feature of dancers' costumes from ancient times in China, this overlong sleeve is attached to the coats of actors in some female roles. In the hands of a skilled actor, it is possible to inject great artistry into shaking down the flowing sleeve to fit the length of the arm and show the hand. A particular master in the arts of the following sleeve was one of the "four great famous dan", Cheng Yanqiu, who designed several gestures to manipulate it more gracefully.❶

拙译　尽管京剧中的不同人物角色，从文人到青年女性或贫穷妇人，其着装不如长袍或盔甲精美，但也都从不同角度体现出京剧服饰的复杂性。尽管衣着颜色不同，但是蓝色一般是贫寒女性的专属。术语青衣（青衣这个角色一般指具有优雅气质的女性，走路轻快，迈着碎步），指代不一定贫穷，但是比较娴静的女性角色。京剧戏服源自我国古代，袖口上大多缝接着一块白色的长纺绸，被称为水袖，最初长度为33~40厘米，然而旦角行当中青衣的水袖最长已达到73厘米以上。一般而言，水袖长短要以演员的个子高低和手臂长短而定，要方便演员通过水袖的表演来展示舞美艺术动作，同时水袖对于塑造女性角色尤为重要。由于京剧表演艺术的不断发展，水袖的表演也有了很大的变化。抖拂的轻重快慢、姿势的大小高低，都随着角色的心理活动而各不相同。这既有助于表现角色的身份、性格和感情，又可以增强舞蹈的美感。"四大名旦"之一的程砚秋掌握了水袖的表演技巧，表演功夫精深，并设计总结了好几套动作，使水袖表演更加优雅自如。

原文　The costumes have become one of the defining features of the traditional items based on old stories. The term Guzhuangxi, which means literally "ancient costume drama" has come to refer to any items that treat traditional themes. In the Peking Operas on contemporary themes, the costumes and headgear are very much more realistic and simple than in the traditional pieces. In these, the appearance and the color of the dress clearly show not only the character's role but the period, season and context of the scene itself.❷

拙译　基于老故事，戏装已成为传统京剧剧目的特征之一。术语"古装戏"的字

❶ MACKERRAS C P. Chinese Theater：from Its Origins to the Present Day［M］. Honolulu：University of Hawaii Press，1983：47.
❷ 同❶45–46.

面意思是"穿着古代服装表演的戏剧",用来指任何与传统主题相关的内容。在以当代为主题的京剧艺术中,其服饰和头饰比传统京剧中的要更客观、更简洁。实际上,京剧服饰和头饰的外观与颜色不仅清楚地揭示了人物角色,而且暗示了剧情本身发生的时间段、季节和环境。

原文 京剧中的大部分服装在很多戏里都能通用。比如,一套皇帝的服装在这个戏里皇帝的角色能用,可能在其他皇帝的戏中也可以用;一套秀才穿的服装,无论哪个戏里的秀才角色都可以穿着。京剧的服装是以古代日常生活的服装为基础的,且在经过很多艺术家的提炼、概括、美化和装饰后,形成了整套类型化,或者说程式化的专用服装。这些服装的概括能力相当强,使用的范围也很广。京剧服装不仅是帮助演员塑造角色外形的有效途径,更是帮助演员抒发内心感情、刻画心理活动、揭示精神世界的有力手段。❶

译文 Most costumes in Peking Opera can be worn in many different plays. For example, the costume for an emperor in one play can be worn by emperors in other plays, and the costume for a Xiucai (a scholar in ancient China) can be worn in any play by such a character. Based on the daily clothing worn in the Ming dynasty and earlier, the costumes in Peking Opera are selected, refined, beautified and decorated by generations of artists, resulting in whole set of specialized clothing designed for different characters. The costumes are highly efficient as one article can be used on many occasions. In general, the costume is an effective means for an actor to build a character's outward image, and even more so to express and depict the characters inner emotions and mental activities. ❶

原文 京剧服装的第一个特点,首先就是用最少的服装来适应尽可能多的剧目演出的需要,不必每演一个戏就重新换一套服装。京剧的传统剧目数量众多,大约有数千部,大多取材于历史故事,这就是京剧需要一整套概括能力比较强、使用范围比较广的服装的原因。京剧服装只要表示出人物的身份或者地位就够了,人物的性格、境遇等就都由剧本的文字和演员的表演来展现。❷

译文 The Peking Opera costuming has three distinctive features. The first is its adaptability. That is to say that a costume can be worn in many different plays so that the number of costumes can be kept to the bare minimum. This results from the fact that Peking Opera has a repertoire of several thousand traditional plays which mainly derive from historical stories. It will do for the costume to indicate the identity and social status of a character. As to personality and circumstances, they will be revealed more by the script and performance.❶

---

❶ 孙萍. 中国京剧百部经典英译系列:秋江 [M]. 北京:中国人民大学出版社,2012:160.
❷ 同❶,略有改动.

原文 京剧服装的第二个特点是不受季节的限制，不同于生活中春夏秋冬服装的定规。除非这部戏的剧情有些特殊的要求，才在服装上稍微添加一些装饰性的点缀，使观众意会。京剧中天气的变化都要通过演员的表演表现出剧中人冷暖的感觉，而不是通过服装展现。❶

译文 Second, the costume, unlike regular clothes in daily life, does not have to change with the seasons. Usually the change of weather is hinted by the player's performance, unless in special situations some decorations are added to give a clue to the audience. ❷

原文 京剧服装的第三个特点，也是最重要的特点就是为剧情服务。京剧服装是以京剧表演艺术的需要为原则进行设计的。可以说，京剧舞台艺术的所有组成部分都是为演员的表演而服务的。京剧服装的美学观点和设计原则有利于舞台动作的复杂化，舞蹈动作幅度的加大化，从而延伸舞蹈动作的内涵意义。京剧的舞蹈动作是比较复杂的，京剧服装的设计、制作必须配合和辅助这种舞蹈动作的运用。❸

译文 Third, the costume serves the plot, which is its most important function. It is designed in accordance with the aesthetic principles of Peking Opera. All components of Peking Opera stage art aim to enhance the performance. The aesthetic and designing principles of the costumes is providing variety to stage movements. With the help of costumes, the movements are visualized and emphasized. Thus, more meaning is added to the performance.❸

关于京剧服饰的内容，潘霞凤在她的 *The Stagecraft of Peking Opera* 一书中也有大量介绍，现摘录并整理相关内容，在以英汉双语方式阐释的同时，补充解释个别特色词汇，具体如下。

原文 Costume, generally referring to what an actor or actress wears on the stage, is technically termed in Peking Opera and local operetta as Xingtou or, more popularly, Xiyi in Chinese. The Peking Opera costume can be traced back to the mid-14th century, fashions of people of the Ming dynasty (1368−1644). It underwent changes gradually and continually, and has emerged as it is seen nowadays. In general, the Chinese Opera costume is characterized by the following features.❹

拙译 一般来说，男女演员在舞台上的着装，在京剧和地方小剧中被称作是行

---

❶ 孙萍. 中国京剧百部经典英译系列［M］. 北京：中国人民大学出版社，2012：160，略有改动.

❷ 同❶160.

❸ 同❶161.

❹ PAN XIAFENG. The Stagecraft of Peking Opera［M］. Beijing: New World Press，1995：107.

头，在汉语中更普遍的说法是戏衣。京剧戏衣的历史可以追溯到14世纪中期，明代（1368—1644）人们的时尚中。随后历经不断的变化发展，最终成为今天人们看到的样子。总体而言，京剧戏服具有以下特点。

原文（1）Anachronisms are allowable. A performers' costume primarily designates his or her role on the stage no matter when or where the action takes place. Characters, whether they are ancient of the pre-Christian era Shang or Zhou, or their descendants in pre-modern China, appear on the Chinese Opera stage wearing costumes suitable to their roles. However, in the courses of theatrical development a few items of Manchu-style apparel, for instance, the archer's dress (Jianyi) and the mandarin's jacket (Magua), have been introduced into the ensemble of Peking Opera costume.❶

拙译（1）京剧服饰打破了时空界限。表演者无论何时何地演出，其服装的设计主要取决于其在舞台上的角色定位。无论戏曲角色是生活于我国商代或周朝，还是现代社会之前，在京剧舞台上的人物只关注适合自己角色的服饰。然而，在戏剧发展过程中，一些满族服饰如弓箭衣（箭衣）和汉族人穿的上衣（马褂）也都被纳入了京剧服饰系列组合中。

原文（2）Regardless of the four seasons, the opera costume is the same, although the elements are described in every scene and have to be made clear by the actor's movements. Thus, Chen Daguan, a male lead in story of the *Number One scholar*, wears a patched garment (Fuguiyi) in spring, and his counterpart, Mo Ji in *Jin Yunu*, wears a similar one in the dead of winter.❷

拙译（2）京剧服饰的选用打破了四季界限。尽管季节在每个场景中都有描述，但依然需要借助演员的表演动作才得以清楚体现。正如《状元谱》中的男主角陈大官，春天里身着拼布衣（富贵衣）；而《金玉奴》中的书生莫稽，即使是寒冬腊月，也依然穿着类似的戏服。

原文（3）Peking Opera costume has to distinguish a character's sex and status at first glance. Usually, an emperor shows up on the stage in a nine-dragon crown (Jiulongguan) or imperial crown (Wangmao) and a yellow informal robe (Huangpei❸) or ceremonial robe (Huangmang). Officials are dressed differently in accordance with their ranks in the bureaucratic hierarchy. For example, a prime minister often takes as his formal attire a decorated chancellor's hat (Xiangdiao) and a colord robe (Xiangsemang), while a magistrate will put on a black gauze hat (Wushamao) and an official robe (Guanyi). As for the commoners, they seem to be eligible only

---

❶ PAN XIAFENG. The Stagecraft of Peking Opera [M]. Beijing: New World Press, 1995: 107.

❷ 同❶107–108.

❸ 本书中有关"帔"的英译，有两种形态：Pi（传统戏曲服装术语）和Pei（为尊重原著保留原文）。

for drab clothes (Laodouyi), black-cuffed jackets (Chayi) and commoner's coats (Xizi❶). A young lady of noble birth may be privileged to have a sort of robe that is termed Pei in Peking Opera costume, and a handmaid will only have access to blouse and trousers (Ao and Ku), articles which used to be worn by common people in daily life. In terms of symbolism, Peking Opera costumes may well be regarded as having the main function of marking off people from all walks of life, be it noble or humble, civil or military, as well as in or out of office.❷

拙译 （3）依据京剧表演者的戏服，观众一眼就能区分出不同角色的性别和地位。因为通常皇帝在舞台上是头戴九龙王冠（九龙冠）或皇冠（王帽）、身穿非正式的黄袍（黄帔）或正式的长袍（黄蟒）。官员们依据封建官僚体制下的官衔级别差异，穿着也各不相同。例如：大臣经常戴着装饰性的礼帽（相貂），穿着彩色的长袍（相色蟒）；文职官员则头戴黑色纱帽（乌纱帽），身着官袍（官衣）。对于平民百姓而言，他们似乎只有资格穿着单调的褐色衣服（老斗衣）、黑布衣（茶衣）和平民外套（褶子）。贵族出身的女性有资格穿用京剧中的袍衫——帔，女仆则只能穿着袄和裤子，这些也是百姓生活中的日常着装。从象征意义而言，京剧戏服的主要功能是区别社会中的不同行业、不同阶层，贵族或平民、文官或武官，以及在职与否。

原文 （4）By means of a subtle symbol, opera costumes may give expression to sharp distinctions between the good and the evil or, preferably, the loyal and the wicked. The wings (Chizi) attached to a gauze hat indicate a loyal officials if they are ob-longish trapezoidal ones, such as the ones Mao Peng wears in *Four Successful Candidates*. In contrast, Gu Du, a corrupt officials, is made to wear a gauze hat with rhomboidal wings, and Liu Ti, a fatuous magistrate, wears a hat with round wings.❸

拙译 （4）借助微妙的象征意义，京剧服饰使善与恶之间的界限分明，或者说忠诚与邪恶的区别明显。纱帽中的方翅纱帽（翅子）暗示尽职的官员，如《四进士》中毛朋的穿着。相比之下，贪官顾读，戴着菱形翅子的纱帽；而昏官刘题，则戴着圆形翅子的纱帽。

补译 舞台上各种上层人物所戴的冠帽成为区别他们身份的明显标志。官帽上翅子的样式基本分为三类，即向上、平直和向下。向上的称为朝天翅，如皇帽、皇巾、九龙冠、扎镫、相巾等冠帽，只有皇帝、重臣和高级官员才有资格戴。一般文官戴的帽子是方翅纱帽、尖翅纱帽、圆翅纱帽或桃叶翅纱帽，都用比较平直的翅子。宰相戴的相貂虽然也用了平直的翅子，但其翅子很长，有别于一般官员的帽子。其他非官员或官员在私下戴的各种巾帽则都用向下的翅子或飘带。

❶ 本书中有关"褶子"一词的英译，有两种形态：Xizi（传统戏曲服装术语）；Xuezi（为尊重原著保留原文）。
❷ PAN XIAFENG. The Stagecraft of Peking Opera［M］. Beijing: New World Press, 1995: 108.
❸ 同❷108-109.

官帽上的翅子也是对官员褒贬的写照。舞台上的方翅纱帽一般为生角所扮的正直官员所戴，故称忠纱；尖翅纱帽则多为净角所扮的奸诈官员所戴，故称奸纱；丑角扮的贪官污吏，则戴圆翅纱帽，其翅子的花纹有的直接是个铜钱，象征其鱼肉人民，所以这种翅子也被称为金钱翅。

拙译 The hats worn by all kinds of upper class people on the stage are obvious signs to distinguish their identities. The patterns of the wings (Chizi) on the official hat can be divided into three categories: upward, straight and downward. The upward one is called the wing (Chizi) to the sky, such as the imperial hat, imperial kerchief, Jiulong (decorated with nine dragons) crown, stirrup-shaped crown, phase kerchief and other crowns, which are only fit for the emperor, close courtiers and senior officials. The hats worn by civil servants are square wing gauze, pointed wing gauze, round wing gauze or peach leaf wing gauze, which are all with flat wings. Although the prime minister's mink also uses straight wings, its wings are very long and are different from the others. Usually people may choose all kinds of hats and scarves in their daily life, such as the hats with downward fins or streamers.

The wings on the official hat also reflect the praise and criticism to the officials. The square wing gauze hats on the stage are usually worn by the officials of integrity, so they are called "loyal gauze". The sharp winged gauze hats are often worn by treacherous court officials, and they are called as "Jiansha" in Chinese pronunciation. And the corrupt official disguised as a clown usually wears a round wing gauze hat, even with the copper coin patterns on his wings, symbolizing their excessive exploitation of people, so this kind of wing (Chizi) can also be called as "money wing".

原文 (5) Accessories, though apparently of little or no account as compared with such principal items as crowns and robes, many nevertheless function to bring about more dramatic effect on the stage, for instance, the wings attached to a hat, the plumes (Lingzi) pinned to a helmet and the cascading sleeves (Shuixiu) sewn to a garment, even if they have never had any practical use.❶

拙译 (5) 配饰虽小，与诸如冠和袍子这类的主要服饰相比实在是微不足道。如帽子上的翅子、头盔上的翎子、衣服上的水袖，尽管没有实际用途，但是这些配饰对于舞台戏剧效果具有重大的现实意义。

补译 水袖是演员在舞台上情绪放大或延长的手势。舞动水袖是我国京剧的特技之一。演员利用服装袖口上缝接的一段白绸（水袖），展示丰富的、舞蹈化的动作以表达人物心理。水袖舞蹈的身段和动作姿态多达数百种，基本动作包括勾、挑、撑、冲、拨、扬、掸、甩、打、抖等。

---

❶ PAN XIAFENG. The Stagecraft of Peking Opera [M]. Beijing: New World Press, 1995: 109.

拙译 The Beijing Opera features Shuixiu (water sleeve dance), a gesture that artists use to visualize the character's emotional world, as a distinct feat. Waving loose white silk sleeves attached to cuffs, experienced artists portray the character's psyche through the dancing movements. There are hundreds of gestures and dancing movements of Shuixiu. The elementary movements embrace Gou, Tiao, Cheng, Chong, Bo, Yang, Dan, Shuai, Da and Dou in Chinese pronunciation and so on.

原文 By making one of the wings stir while at the same time keeping the other still, a performer in a black gauze hat may suggest that he is cudgeling his brains for ideas. And, in the end, when the two wings are in motion simultaneously, they are betraying his satisfaction at having found something new.❶

拙译 通过一个帽翅的舞动和另一个帽翅的保持不动，折射出表演者正在冥思苦想绞尽脑汁；而当两个帽翅同时舞动时，则流露出表演者有新发现的得意与喜悦。

补译 耍纱帽翅属于京剧艺术身段的特技表演。演员耍动帽翅时需要以颈部为轴心来带动纱帽翅上下摆动、左右旋转或前后绕圆圈。有时只单翼抖动，有时则双翼同时旋转搅动。耍时帽翅欲停则停，欲动则动，有时则或停或动，充分揭示了人物角色内心复杂的心理活动。

拙译 Stirring the wings of the gauze cap (worn by feudal officials) is one of the stunt performances of the Peking Opera to the actors, to stretch their necks as the axis to drive the wings of the gauze cap up and down, left and right or to circle around. And sometimes only a single wing is shaken, or the two wings are rotated and agitated at the same time. whether to stir or stop the shaking, it can be very well, managed, which fully reveals the complicated psychological activities of the characters in the opera.

第三节
传统京剧服饰特点与文化内涵

（一）传统京剧服饰特点概述

京剧服饰具有一套完整且动态的体系。在京剧服饰发展过程中，随着社会审美的

---

❶ PAN XIAFENG. The Stagecraft of Peking Opera [M]. Beijing：New World Press，1995：109.

变化而进行改进与创新，从而使整个京剧服饰体系变得充实且新颖。但在这个过程中，京剧服饰的基本特点及遵循的审美原则则未曾改变。京剧服饰具有艺术性，因而随着朝代、季节、个人审美的不同产生了服装差异，自成体系。京剧服饰的特点可概括为：符号性、程式性、可舞性和色彩丰富等。

京剧服饰的"符号性"在于其"符号"所具有的能指与所指及其关系特征。京剧服饰的物质形态，如款式、颜色、纹样、质地等因素构成了京剧服饰的能指，而这些具有能指功能的特征因素通过演员在京剧表演中所呈现的戏剧角色，包括其蕴含的宗教、民俗、审美和道德等意义，构成了所指。同时，京剧服饰符号性特征本身也是基于人的主导地位而产生的。京剧服饰的"程式性"在于其几乎所有人物的着装都具有程式性，基于人物身份特征与所处环境条件等，找到与其吻合的"位置"，遵循"宁穿破，不穿错"的原则，提供相对应的服饰。京剧服饰的"可舞性"在于强化服饰艺术的造型功能，借助耍水袖、踢褶子等舞蹈动作，凸显演员肢体语言的技术性与舞蹈性，提升观众的视觉愉悦度，如中国戏剧的翎子、水袖等就是戏剧舞蹈发展的产物，而中国戏剧独特的翎子舞、水袖舞等表演则又是戏剧服饰可舞性进一步促成的结果。

京剧服饰的另一大特色是色彩的鲜明性。京剧服饰通过"色彩斑斓"或瑰丽浓重与简约淡雅的色彩对比，使浓厚艳丽与清雅文静相辅相成，成功塑造舞台人物形象，彰显人物特色。上述四大特点既是我国传统京剧文化艺术的结晶，也是我国古代戏剧演出实践的必然产物。

（二）京剧服饰颜色的文化内涵

京剧服饰运用色彩来表达京剧人物之间的关系及其思想内涵，不同身份定位的人物服装使用特定的色彩，这些色彩的选用标准则源于中国古代传统色彩观念。中国先哲十分注重自然界中的色彩，他们将色彩与组成世界的元素相互联系，引发系列哲学思考。即将"五行"思想与"五色"观相联，将"金、木、水、火、土"看作是构成世界的五种基本元素，用"五行"解释宇宙生存及系统运转。"五行"与"五色"相配，"金、木、水、火、土"对应"白、青、黑、赤、黄"。

红、绿、黄、白、黑为上五色，倍受尊重。红色指的是色彩中的大红，而非色光中的品红色，也不是原色（一次色）。在京剧服饰中，身份地位较高的角色服红色，新人婚礼中也常用红色，代表喜庆与团圆。绿色指的是原色中蓝和黄相加的间色（二次色），也称老绿色，而非色光中的绿色。在京剧中，服绿色代表此人气质神勇。黄色指的是明黄，类似于颜料中的中黄，属间色，而非色光中的黄色，在京剧中黄色象征人物身份尊贵、性格温柔。白色属于中性色（调和色），代表角色潇洒英武、品格高洁，此外在表现角色服孝的场景时也使用白色。黑色属于中性色（调和色），黑古称玄色，在戏剧中称为青，黑色象征对天的崇敬，既不同于古代色彩称谓中的青（蓝色），也不同于现代色彩概念中的青（蓝色），在民族审美意识中给人以庄重气派的感觉。在京剧中，黑色代表性格粗犷、气质威武。以紫、粉、蓝、湖、香为主的下五色，相对于上五色地位较低。紫色代表老龄；粉色代表年轻英俊；蓝色代表官职较低；湖色代表性格文静，气质儒雅；香色代表老龄，性格稳重练达。笔者摘录美国学者关于我国京剧传统色彩的英文阐释，并将拙译一并呈现如下。

原文　The upper colors[1] are red (Hong), green (Lü), yellow (Huang), white (Bai) and black (Hei), which correspond with the colors of the five directions. As the upper colors are the purest, and considered the finest in historical terms, they appear on garments of nobles and officials at court in their formal dress and on generals, and commonly appear more on male and leading characters. Their designation be as "upper colors", however, it refers to their frequency of use onstage, rather than exclusively to status. The sequence of the colors also indicates the order in which they are stored, rather than rank. [2]

拙译　上五色，指红、绿、黄、白、黑，五种色彩对应着五个方位。上五色是最纯净的颜色，也是历史上公认的最好的颜色，是贵族或官员上朝时所穿正装的颜色，通常是男性或主要人物的衣服色。这些颜色在京剧中被称为上色，主要是基于其在舞台上的使用频率，而非因为这些颜色的特有地位。颜色的顺序暗指它们的存储顺序，而不是色彩等级。

原文　The hues of the lower colors vary somewhat, but they can generally be listed as purple (Zi), pink (Fen), blue (Lan), lake blue (Hulan), which is similar to sky blue or turquoise, and bronze or olive green (Qiuxiang or fragrant autumn). The purple color Zi resembles maroon in the western sense and will be called maroon for clarity. The lower colors are generally worn in informal scenes, much as they were in the past, though maroon and blue can be used for court, as well. As most of the lower colors are not primaries, they also have more variation in execution. Several pastel versions of blue fall into the lake-blue category, such as aqua and turquoise, and beige and mustard can also be ascribed to bronze or olive green. In addition to their historical and cosmic meanings, the colors attributed to the upper and lower categories contain another layer of connotations for personality traits.

Though the five upper and five lower colors are the principal colors that are drawn on when constructing costumes, no colors are forbidden onstage. Those that fall outside the ten tend not

---

[1] 本书中有关"上、下五色"一词的英译，有几种形态：the upper colors, the lower colors, the five upper and five lower colors, Upper Five Colors 等，为尊重原著保留原文。

[2] BONDS ALEXANDRA B. Beijing Opera Costumes: the Visual Communication of Character and Culture［M］. Honolulu: University of Hawaii Press，2008：72.

to be used as much simply because of the preference for "talking colors" that project a stronger meaning. A third category contains other colors that appear onstage, the complicated or mixed colors (Zase) including orange, gray and brown. All of the colors in this category have to be mixed using two, three or more colors. While every color has a meaning, the colors in this group have the least significant meaning as well as the lowest position in the hierarchy. Neutralized fabric colors are rare on the dress of the principals because of the desire for meaningful, as well as attractive colors. Instead, light browns and beiges, called tea colors (Cha), are used for the costumes of servants, workers and the elderly. Gray, off-white and other neutral colors appear on monks, as they do in actually. ❶

拙译 下五色的色调稍微有所不同，通常是紫色、粉红色、蓝色、类似天蓝色或绿松石的湖蓝色、青铜色或橄榄绿的颜色（秋香色或香秋色）。紫色类似于西方的栗色，因其颜色清晰而被称为栗色。下五色在过去一般用于非正式场合，尽管栗色和蓝色也被用作于宫廷色。因为多数的下五色不是基础色，而且在使用中有很多变数。几种柔和的蓝色可列入湖蓝色系，如浅淡的蓝绿色和绿松石色；米色和芥末色也可归入青铜色或橄榄绿色。除去在时间和空间的含义之外，这些颜色因为上五色和下五色的分类，还包含了另一层的具有个性特征的内涵。

虽然上、下五色都是服装运用的基本色调，但是舞台上没有禁忌色。那些没被算在这十种颜色之内的色调往往不被轻易使用，因为对于这些"会说话颜色"的偏爱蕴含着更多的意味。第三类是在舞台上出现的其他颜色，即混杂的颜色（杂色），包括橙色、灰色和棕色。这一系列中的所有颜色都是两种、三种或更多种颜色的混合。然而每种颜色都有各自的内涵，至少意味着等级制度中社会的最底层。中性面料颜色很少出现在服饰的基本色调中，因为颜色需要吸引人，需要传递意义。相反，浅棕色和米色被称作是茶色，这种颜色是仆人、员工以及长者的着装色；灰色、米白色以及其他中性色是出家人的着装色。日常生活中的实际情况也确实如此。

原文 Meanings of the Upper Five Colors. A vast range of meanings can be expressed by a single color. A red Mang (court robe) has different significance depending on its embroidery patterns and different connotations from a red Pi (formal robe). Where colors have multiple messages, the color needs to be viewed in context with the other aspects of the garment's form and texture, along with the character's makeup and headdress, and the specific character and scene for the meaning to become more evident. The following sections describe some of the typical applications of color meanings. ❷

❶ BONDS ALEXANDRA B. Beijing Opera Costumes: the Visual Communication of Character and Culture ［M］. Honolulu: University of Hawaii Press, 2008: 72.

❷ 同❶73.

拙译 上五色包含着不同含义。这一系列颜色的不同含义可以通过单一色彩来表达。红色蟒袍（宫廷袍）基于刺绣图案或红色帔（正式袍）的差异，内涵也各不相同。在颜色传递多重意义时，颜色的内涵需要结合具体服饰的款式、质地，包括人物角色的妆容、头饰以及特定的角色和场景一并考量，只有这样才会更有说服力。下面的这部分描述了一些有代表性颜色的典型应用。

原文 Red carries the favorable associations of respect, honor and loyalty connected with essentially good characters. Red Mang are often worn at court by the high-ranking nobles and principal statesmen who are the steadfast subjects of the emperor, and second in rank only to the emperor. Red also appears at court when worn by the highest-ranking officials, who are attired in red Guanyi (official robe). Red can reflect passion at different levels, for it is worn by executioners and criminals, and represents the deprivation of life. Men of power, intensity, and status who are not necessarily on the side of good can also appear in red garments. In contrast, the Nümang (female court robe), Gongzhuang (palace garment), and the Nükao (female armor) are often constructed from red fabrics because red is considered the most beautiful of the colors. Since red was worn for weddings in real life, red also appears onstage for marital unions and can be worn by matchmakers. ❶

拙译 红色承载着尊重、光荣和忠诚，总是与正面角色相联系。红蟒在宫廷中是高官贵族，以及那些在朝廷中绝对忠实于皇帝的大臣们的专属，他们的官位仅次于皇帝。红色也出现在朝廷官位最高的官吏身上，作为他们的官服（官衣）颜色。红色可以反映不同层次的情绪，可以为凶手和罪犯所穿戴，代表着剥夺生命。具有权力、威势和地位的男性，但不一定是好人的角色同样也可以穿红色衣服。相形之下，女蟒（女宫廷服）、宫装（官廷服装）以及女靠（女盔甲）通常也由红色面料裁制，这是因为红色被认为是最美的颜色。在现实生活中红色也常用于婚礼，在舞台上红色也是婚礼服，包括红娘也穿红色。

原文 Green indicates a high-ranking or military function for the wearer and is worn by generals on and off the battlefield and civil officials in charge of military affairs. Green may also be worn by princes or regents. A slightly sharper color of green appears on the Xuezi (informal robe) worn by Chou (clowns), and with a shift in color, the meaning changes to indicate low-minded and petty characters who may have fraudulent ambitions. Another color of green in the skirt (Chenqun) on an older woman indicates the worthiness of old age and maturity. ❶

拙译 绿色暗指穿着者的高级别或军人身份，因为掌管军事事务的人无论在战场上或官场上，包括处理民事的官员都穿绿色。王子或摄政王也都穿绿色；而京剧中小丑（丑角）穿着的褶子（非正式长袍）是稍微艳丽的绿色。随着颜色的变化，这种服

---

❶ BONDS ALEXANDRA B. Beijing Opera Costumes: the Visual Communication of Character and Culture ［M］. Honolulu: University of Hawaii Press，2008：73.

京剧服饰文化传播及其文化内涵

第二章

27

饰就成为卑鄙狡诈、野心勃勃的小人物的象征。老妇人穿着的绿裙子（衬裙）则用来暗指女性的高龄和成熟。

原文 Only the emperor, members of the imperial family, and their retinues can wear a specific hue of yellow, and this color choice comes from the historical designation of yellow for royalty. Although yellow represents the most powerful color, it is listed as third and placed in the storage trunks third so that the garments do not "get a swelled head". Characters with wisdom, who are worthy of respect, who can solve problems cleverly and outwit their adversaries, may wear other shades of yellow. Sun Wukong, the Monkey King, wears yellow because he has been appointed a saint of heaven and a Buddha of triumph, and the color of heavenly saints is yellow.❶

拙译 黄色是只有皇帝、皇室成员和他们的随从才能穿戴的一种特殊色调，而这种颜色的选择源于黄色在历史上被赋予的王权含义。然而，黄色虽然代表了最强权的色彩，但它被列为第三名，所以黄色的服装在衣柜中位居第三，不会"得意忘形"。智者、受人尊敬者、能巧妙解决问题者、战胜对手者，在京剧中可能都会穿着某种黄色。如猴王孙悟空因被任命为天国的圣人和胜利之佛而穿黄色，天上圣徒的衣服颜色也是黄色。

原文 White garments can be used for men who have grace, charm and loyalty to their country. Both Xiaosheng (young men) and Laosheng (mature men) are seen in white Mang and white Kao (armor). White garments can also be used to show the youth of female characters, as Qingyi (young to middle aged woman) can wear white Nüpi (woman's formal robe) and all of them wear white pleated skirts (Baizhequn) with their other garments. A pleated white "bloated" skirt (Yaobao) worn over the other garments indicates a sickly or pregnant wearer. The tradition of wearing white for mourning has transferred to stage usage. Characters in mourning may appear with a simple strip of white fabric draped over their heads, under their headdresses, or they may dress completely in white.❶

拙译 白色服装适用于对祖国忠诚、优雅且有魅力的男士。小生（青年男子）和老生（成熟男子）都穿白色蟒和白色靠。白色服装也适用于年轻女性，如青衣（年轻或中年女子）可以穿白色女帔（女性的正式袍装），而且所有青衣都可以穿着白色的褶皱裙子（百褶裙）。稍显"臃肿"的白色百褶裙（腰包）穿在其他衣服的外面，暗示那些病态或怀孕的青衣。此外，白色服饰表示哀悼的传统已传承到舞台，京剧中表示哀悼的人会把白布条简单搭在头衣的下面，或者干脆全身着白色服饰。

原文 A black Mang represents characters who enforce laws and penalties, and who

❶ BONDS ALEXANDRA B. Beijing Opera Costumes: the Visual Communication of Character and Culture［M］. Honolulu: University of Hawaii Press, 2008: 73.

are straightforward, brave, honest and upright. Outside of court wear, black garments take on many other meanings that are quite opposed to the elevated significance of the symbolism and application of colour court. For Qingyi, a black center-front-closing Nüxuezi (woman's informal robe) indicates poverty that may have been brought about through the loss of husband or family. The color was originally called Qing (dark or black), a hue unique to Chinese culture that falls somewhere between blue, green and black; it is the shade of the night sky just as it turns dark. The word Qing combined with Yi (clothing) evolved to give the characters who wore such colors their role name, Qingyi. Xiaosheng who have not yet passed their exams dress in black Xuezi, considered one of the lowest garments, because their financial resources are precarious. Servants wear black, from the heads of household to waiters. Stealthy fighters who are illegal or mysterious and want to escape recognition also wear black.❶

拙译 黑蟒代表直率、勇敢、诚实和正直的执法人或行刑的人。除去宫廷穿用之外，黑色服装还蕴含了许多其他意义，这些含义与宫廷色的应用与象征意义大不相同。因为青衣穿的这种黑色对襟的女褶子（女性非正装），暗示了因为丧夫或家里有人去世了而导致的贫穷。这种颜色最初被称为青（深色或黑色），是中国文化中独特的色彩，是介于蓝、绿和黑这三种颜色之间的色调，仿佛夜晚天变黑时夜空的阴影。"青"和"衣"这两个字组合在一起演化为穿这种衣服的人物角色，青衣。小生在中举之前因为经济来源不确定，一般都穿黑褶子，这是身份地位最低级别的标志。如仆人也都穿黑色，从总管到侍者。此外，那些行动诡秘、想要避免被人认出的非法或神秘的打手也都身穿黑色。

原文 Meanings of the Lower Five Colors. Maroon is used for the upper ranks in the Guanyi, and Laosheng wearing this garment are generally noble and forgiving. Maroon Pi/Nüpi suggest an older age and the respect that goes with that status. A Mang in maroon is more likely to be worn by a Jing role, to indicate high-ranking and dominating characters. When there are too many characters assigned to wear red in a scene, one or more may be shifted to maroon without changing the meaning.❷

拙译 下五色的含义。栗色常用于位居高位的官员的官衣，穿栗色的老生一般都较高贵且宽容待人。栗色的帔或女帔暗指受人尊重的长者。栗色蟒通常是京剧中净角的穿着，暗指其地位高和个性霸道。当太多角色在一个场景中被安排穿红色时，在不改变其所指意义的前提下，一个或更多的角色就要换成栗色的装束。

原文 Pink indicates youth in traditional Jingju. A pink costume on a Xiaosheng indicates he

❶ BONDS ALEXANDRA B. Beijing Opera Costumes: the Visual Communication of Character and Culture [M]. Honolulu: University of Hawaii Press, 2008: 73–77.

❷ 同❶77.

is young, handsome and romantically inclined. Young Qingyi often wear Nüpi in delicate pink, and it is the second most common color of the Nükao.❶

拙译 粉红色在传统京剧中代表青春。小生的粉红色装束暗示他很年轻、英俊且浪漫。年轻的青衣也经常穿嫩粉色的女帔，这是女靠中第二大常用的颜色。

原文 The third of the lower color is a royal blue. A blue Mang signifies high status and may be assigned to virtuous characters who are calm and firm. Another use for blue Mang is for characters with military duties in charge of justice and punishment. Extremely brave and fierce warriors wear blue Kao. Blue also carries a significance of rank, with a blue Guanyi being worn by those in the lowest ranks. Blue relates to youth in the cosmic order, but both youthful and loyal court officers and generals can be dressed in blue-trimmed white garments. In domestic situations, mature couples may be dressed in matching blue Pi. Blue and green, because of their close association in Chinese color theory, are parallel in their meanings onstage, with blue garments being a slightly less powerful version of green ones. When more than one character onstage needs to wear green, the lesser one may be switched to blue. Unembroidered blue satin Jianyi (archer's robes) appear on lowly soldiers or servants, and cotton garments in blue colors similar to the indigo dyes used in real life, are worn by merchants and servants.❷

拙译 下五色中的第三种颜色是贵族蓝。蓝色的蟒暗指具有较高社会地位，平和坚定且生性善良的人。蓝蟒也是军队里的服役者、主持正义者和惩戒者的专用色，尤其指那些勇敢强悍的武士。蓝色同时也蕴含着官衔的大小，蓝色官衣为最低级别官员穿用。此外，蓝色在宇宙秩序中意味着年轻，但是年少的宫廷官员以及众臣只可以穿着镶有蓝边的白色官服。在国内，到一定年龄的夫妇还可以配以蓝帔。在我国色彩理论中，蓝色和绿色的关联度高，因此在舞台上的蕴意也近乎一致，只是蓝色比绿色在权力意义上略逊一筹。当舞台上不止一个角色需要穿绿色时，较小的角色可能会调整为穿蓝色。无绣花的蓝色缎面箭衣（弓箭手的长袍）是地位卑微的士兵或仆人的专属，类似于现实生活中商人和仆人穿用的靛蓝色棉服。

原文 Lake blue is utilized for youthful roles, both male and female. The Xiaosheng wearing Xuezi in one of these lighter blues are considered scholarly, elegant and handsome young men, and their Qingyi lovers are quietly demure in their lake blue Nüpi. These blues most often occur in clothing worn at home, rather than at court.❷

拙译 湖蓝色用于京剧中年轻的男女角色。着装为淡蓝色褶子的小生意味着有学

---

❶ BONDS ALEXANDRA B. Beijing Opera Costumes: the Visual Communication of Character and Culture［M］. Honolulu: University of Hawaii Press，2008：77-79.

❷ 同❶79.

问、有风度，英俊潇洒，他们的青衣恋人则身着蓝色女帔，端庄娴静。但是，这些蓝色经常出现在居家服中，而非宫廷所用的正装。

原文 The fifth of the lower colors translates variously as olive green, bronze, gray, brown, beige and copper, and are all part of a range of colors that are used to represent elderly characters in traditional Jingju. At court, both men and women of rank can wear an olive-green or beige Mang. Matching olive-green Pi may be worn in private situations by an older couple of higher status, while lower-status mature characters may be dressed in plain Xuezi in olive green or any of the other drab colors.[1]

拙译 下五色中排名第五位的橄榄色被阐释成各种不同的绿色，如橄榄绿、古铜绿、灰绿色、棕绿色、米黄色和铜绿色，并且在传统京剧中所有这些绿色都用来代表年长者的角色。宫廷里职位高低相同的男女都可以穿着橄榄绿或米黄色蟒。地位较高的年长夫妇在非公众场合多穿着橄榄绿帔，低职位的年长者则可能穿着朴素的橄榄绿褶子或其他单调颜色的服饰。

原文 Some colors outside the basic ten can have a meaning ascribed to them by connecting them with similar colors. For example, a range of pastels, including lavender and pale green, is often worn by younger characters and project the same impressions as lake blue. Characters who appear in more than one garment is a play sometimes have both in the same color, particularly for officials. A general in a green Mang at court may change to green Kao to go to battle. If an official character appears in more than one play, he or she may wear the same color in subsequent appearances. A character's clothing color may also come from their name or from the literature written about them. For example, a person with a family name of Huang, meaning yellow, may be dressed in yellow costumes, though not the imperial yellow, and the white snake, Bai Shuzhen, wears only white.[1]

拙译 一些十种基本色之外的颜色的蕴意则归因于与它们关联的近似色。例如，系列粉彩，包括一般属于年轻人穿用的淡紫色和淡绿色，给人以类似湖蓝色的印象。在一场戏中，如果表演者需要穿多件衣服，特别是官员的角色，有时可能会都穿同一颜色。在宫廷中穿绿色蟒的官员可能会在战场上换成绿色靠。如果一位官员的角色出现在多个场景中，他或她可能在随后的出场中穿同色戏装。一个角色的衣服颜色也可能与他们的名字或相关的文字描述有关。例如，一位姓黄的人可能身着黄色服饰，尽管这种黄色有别于帝王黄，而白蛇白素贞则只穿白色。

---

[1] BONDS ALEXANDRA B. Beijing Opera Costumes: the Visual Communication of Character and Culture [M]. Honolulu: University of Hawaii Press, 2008: 79-81.

京剧服饰文化传播及其文化内涵

第二章

在遵循上述英汉对照表述的十种基本颜色准则的同时，京剧服饰还会借助色彩来表现人物的年龄、地位和思想内涵。如京剧中的包拯、张飞和项羽等角色常穿黑色蟒袍。黑色蟒袍不仅体现了角色刚直豪放的性格，也给人以庄重气派的感觉。杨延昭、赵云、周瑜等人物都穿白色，给观众以圣洁、潇洒和儒雅的美感。青箭衣多暗示夜间行动，可以烘托气氛。皇帝穿正黄色，王爵、太子穿杏黄色，元老穿香色或白色，侯爵穿红色。红色既可以用于喜庆的场面，同时也用于临刑的犯人，罪衣与罪裤均为红色，监斩官也要穿红官衣、红斗篷，这些都是古代风俗特点的反映。无论是上五色还是下五色，都使观众可以借助服饰颜色将京剧中人物的角色定位，"一览无余，尽收眼底"。

（三）京剧中不同人物角色与服饰色彩

京剧中服饰色彩在遵循上五色和下五色的同时，注重观众的心理需求，小生、花旦等年轻角色常用鲜艳、明亮的颜色；剧情中喜庆场面用红色，丧悼场面用白色，严格遵从风俗习惯的要求；侠客夜行选用黑色服饰，隐蔽目标，强化黑夜的意境。不同角色的服饰犹如人物性格和心情的晴雨表，无时无刻不在叙述着服饰的"颜"外之意。中外有关京剧人物与服饰色彩的篇章可谓浓墨重彩，故本节从英汉双语不同的表达入手，在汉语译文部分会适度对京剧服饰的内容加以补充与扩展，方便读者理解，其具体内容如下。

原文 Laosheng usually wear upper colors because they are generally scholars, officials and advisors to the throne. Their Mang and Kao are most often red, white or green. For older Laosheng, the Mang may be olive green or beige. When traveling, Laosheng may to wear a red Jianyi (archer's robe) with a black Magua (riding jacket). Laosheng in official positions tend to wear the Guanyi of the higher ranks, in either red or maroon, although supporting Laosheng characters are also seen in blue. When a civil Laosheng dresses for noncourt events, he wears a maroon, olive-green or blue Pi, colors of maturity and respect. Military characters in private quarters can be dressed in a Kaichang in white or red. A Laosheng character in a position of lesser status or wealth may be dressed in a plain Xuezi in a neutral beige or brown, blue olive green or black.[1]

拙译 老生常穿上五色衣服，因为他们一般都是秀才、官员和君主的谋臣。他们的蟒和靠大多是红色、白色或绿色。对于年长的老生，蟒可能会是橄榄绿或米黄色。旅途中的老生可以穿红色箭衣（弓箭手的袍装）配黑色马褂（骑装）。处于官员身份的老生趋向于穿红色或栗色的高级别官衣，尽管次要的老生人物形象也穿蓝色。当文老生无须宫廷事务着装时，就会选择栗色、橄榄绿或蓝色的帔，这些颜色都属于受人尊

---

❶ BONDS ALEXANDRA B. Beijing Opera Costumes: the Visual Communication of Character and Culture［M］. Honolulu：University of Hawaii Press，2008：82.

敬的颜色。战争中的人物在私人空间里可以穿白色或红色的开氅。处于次要地位或没那么富有的老生可能穿朴素的褶子，其颜色是中性的米黄色、棕色、蓝色、橄榄绿或黑色。

原文 Xiaosheng wear the same colors of Mang as the older men: white, red and green, with the addition of pink, a frequent choice. Most Xiaosheng who have a rank at court wear the red Guanyi. A Huaxuezi in lake blue, lavender or pink is the most common garment for Xiaosheng to wear outside of court, when they are well-to-do scholars. If they have yet to pass their exams or if their fortunes have failed, then they wear a black Xuezi, either with a white contracting neckband or a matching neckband, and either plain or with minimal embroidery. Xiaosheng rarely wear a Pi, but when they do, their Pi comes in either yellow for the character of emperor or red for a wedding or achieving the position of top scholar in the examinations.❶

拙译 小生和老生穿着同样颜色的蟒，白色、红色和绿色，偶尔也穿粉色。大多有官衔的小生在朝廷穿红色官衣，在不去上朝、以家境优渥的书生身份出现时，最常穿着湖蓝色、淡紫色或粉色的花褶子。如果他们还没有通过考试或遭遇了家道败落，那么他们则穿黑色的褶子，并选用一个对比明显的白色领口或颜色相配的领口，朴素或几乎不加任何装饰。小生很少穿帔，但是一旦穿帔，其颜色要么是象征帝王身份的黄色，要么就是结婚或状元穿用的红色。

原文 When dressed in Kao, Wuxiaosheng (martial young men) and Wusheng (martial men) are frequently clad in white. Their Kao often have blue borders around the white garments. These young generals may also wear the Kao in red and green. When military young men are in private quarters, their Kaichang are generally white or red. They may also wear a version of the Xuezi made of white satin.❷

拙译 当穿靠时，武小生（习武的男青年）和武生（习武的男子）常会穿着白色抱衣。白色外套的靠经常配有蓝色的镶边。这些年轻的官员也穿红色或绿色的靠。当军队里的年轻男士在非公共场合时，通常穿白色或红色的开氅，但也有时会穿白缎质地的改良褶子。

原文 At court, Laodan wear Nümang in yellow, orange or olive green. The Nüpi for Laodan often comes in olive green, blue or maroon, but it can be brown or dark green as well. If they are still married and appear onstage simultaneously with their husbands, their Pi garments usually

---

❶ BONDS ALEXANDRA B. Beijing Opera Costumes: the Visual Communication of Character and Culture［M］. Honolulu: University of Hawaii Press，2008：82–83.

❷ 同❶83.

京剧服饰文化传播及其文化内涵｜第二章

match those of their husbands. Laodan in lower circumstances will wear the crossover closing Nüxuezi in olive green, gray, brown, maroon or off-white if they are particularly destitute.❶

> 拙译　在宫廷里，老旦穿黄色、橙色或橄榄绿的女蟒。女帔的颜色则是橄榄绿、蓝色或栗色，也可以是棕色或深绿色。有配偶的老旦在舞台上有时会携夫君一起出场，这时老旦所穿帔的颜色和夫君衣服的颜色通常都比较相配。但是如果身处底层，生活贫困，老旦就只能穿交领斜襟的绿色、灰色、棕色、栗色或米白色的女褶子。

原文　When at court, Qingyi wear versions of the Nümang usually in either red or yellow, depending on their position. They may also wear a Gongzhuang with a red bodice and multicolored streamers. In domestic scenes, Qingyi are most likely to appear in Nüpi in colors similar to those worn by Xiaosheng. Their Nüpi come in lake blue, pink, white and yellow, to name a few of the colors. After marriage in a red Nüpi, the Qingyi colors shift to a darker range, often a deep blue or maroon. When they wear the Nüxuezi, they use the center-front-opening version distinct to female characters. The range of colors reflects the pastel range of the Nüpi, unless the Qingyi has lost her position and money, in which case, she wears a black Nüxuezi with a blue border. In more recent performances, this garment has taken on some colors and can be a gray-blue or green, with contrasting border.❶

> 拙译　在宫廷里，青衣根据各自的身份穿红色或黄色的改良女蟒，有时也穿上半身收紧的红色宫装，带有彩色飘带。在家里，青衣多数都穿类似于小生的彩色女帔，具体颜色包括蓝色、粉色、白色和黄色。青衣在身着红色帔完婚之后，帔的颜色会换成比较深的棕色，经常是深蓝或栗色。当穿女褶子时，就选用对襟的款式以别于其他女性角色。色系反映了女帔的柔和程度，除非青衣既没地位、也没钱财，这种情况下的青衣穿带蓝边的黑色褶子。不过在最近的表演中，褶子都会带有颜色，如灰蓝或绿色，镶边都采用对比色。

原文　Qingyi characters are generally dignified, serious and honest. Most of them are faithful wives, loving mothers or chaste, undefiled women in feudal society. The role type got its name from the characters' black dress (Qingyi literally meaning "black dress"). It is also known as Zhengdan (chief Dan) because of its primary position among all the Dan subtypes. The characters are generally young or middle-aged.❷

> 拙译　青衣的人物角色一般都是端庄、严肃且诚实的化身。她们大多是封建社会时忠贞的妻子、慈爱的母亲或纯洁的女性。青衣角色的名字来源于扮演人物的黑色戏

❶ BONDS ALEXANDRA B. Beijing Opera Costumes: the Visual Communication of Character and Culture [M]. Honolulu: University of Hawaii Press, 2008: 83.

❷ YI BIAN. Peking Opera: the Cream of Chinese Culture [M]. Beijing: Foreign Languages Press, 2005: 57.

服（青衣的字面意思是"黑色衣服"）。青衣也被称为正旦（主旦），因其在所有旦角中占有重要地位，一般都是青年或中年人。

原文 Huadan (lively young women) usually have white, pink or blue jackets and skirts (Aoqun), which they sometimes wear with darker vests or collar-and-peplum combinations, or a combination of pastel jacket and trousers (Aoku), with a contrasting apron (Fandan).❶

拙译 花旦（可爱的年轻女性）一般都穿白、灰或蓝的上衣和裙子（袄裙），有时候也穿深色背心或带领的裙子，或者柔软的上衣和裤子的组合（袄裤），配有对比色调的围裙（饭单）。

原文 When Daomadan ("sword and horse" women warriors) go to battle, their Nükao are most often red or pink, with multicolored streamers, but when many women are onstage in armor simultaneously, as in Women Generals of the Yang Family (Yangmennüjiang) , the Nükao comes in a spectacular array of pastel colors. If wearing Gailiangkao (reformed armor), the garments are often blue.❶

拙译 刀马旦上战场时（女武旦），大多穿红色或粉色女靠，并配有多彩的飘带。但当很多女性在舞台上穿盔甲时，如杨家将里的女将们（杨门女将）同时身着颜色柔和的女靠，看上去颇为壮观。如果穿改良靠，衣服则通常为蓝色。

原文 For the Wudan (martial women) in the Zhan'ao and Zhanqun (martial jacket and trousers), the colors are most often red, blue or white.❷

拙译 武旦通常都身着战袄、战裙，颜色为红色、蓝色或白色。

补译 武旦一般是精通武艺的女性角色，主要分为两大类。一类是短打武旦，穿短衣衫。这类的武旦大多不骑马，重在武功，重在说白，还有打出手的特殊技巧，甚至以跌扑取胜，在唱功和表演上不太注重。这样的戏有很多，如《打焦赞》中的杨排风、《泗州城》中的水母、《打店》中的孙二娘、《三岔口》中的店主婆等都属于短打武旦。另一类是长靠武旦，通常也称刀马旦。即妇女也穿上大靠，顶盔贯甲。这样的角色一般都是骑马带刀。

拙译 Female characters that master martial art are called Wudan (martial women). Wudan can be divided into two types: Duanda Wudan and Changkao Wudan. The former gets dressed in short shirt and generally gets no horse to ride. The stress placed on the martial art, recitative, and special skills to fight far outweighs the requirement for actors' singing and acting. There are many characters of this kind, including Yang Paifeng from *the Fighting Jiao*

❶ BONDS ALEXANDRA B. Beijing Opera Costumes: the Visual Communication of Character and Culture［M］. Honolulu：University of Hawaii Press，2008：83，略有改动.

❷ 同❶83.

Zan (*Da Jiao Zan*), Shuimu from *the Sizhoucheng*, Sun Erniang from *Dadian*, and the hostess from *the Crossroads* (*Sanchakou*). Changkao Wudan, also named Daomadan, are required to wear Dakao, helmet and armor and bring broadsword and are permitted to ride.

原文 The Jing wear intense colors to match their bold personalities and to contrast with their face color. At court, they usually wear Mang in red, green, black, maroon or blue. When dressed in Kao for battle, the Jing roles appear quite often in black, red, green, blue and orange. Military Jing usually travel in the Magua, usually black, often over a blue or maroon Jianyi. In personal quarters, Jing often appear in the Kaichang in red, green, black or maroon. They rarely wear the Pi, and if they wear a Xuezi, it may indicate lower status or defeat and can be a dark or neutral color.[1]

拙译 净角通常穿浓烈的色彩以烘托其大胆的个性，并与脸谱颜色形成对比。在宫廷里，他们穿红、绿、黑、栗色或蓝色。当穿着靠上战场时，净角一般穿黑、红、绿、蓝或橙色。行军打仗时，净角一般穿黑色马褂，外罩蓝色或栗色的箭衣；居家时，一般穿红、绿、黑或栗色的开氅。这类角色很少穿帔，要是穿深色或中性色的褶子，就意味着社会地位低下或战败。

补译 京剧中的净，即清洁干净，而净角都是大花脸，看起来很不干净，不干净的反面就是干净，因而名净。

拙译 The Jing mean clean in Chinese,but in actual fact, the "painted face" of the role never looks clean at all. The reason why the role is named as Jing in the opera could be a demonstration of a dramatic contrast.

原文 The Chou seldom have the eminence or personality required for the wearing of the higher status and rank garments. On the rare occasion that they do appear in a Mang, often when they are playing the character of a eunuch, they are dressed in green or olive green. Chou playing eunuchs are also dressed in the Taijianyi (eunuch robe) in yellow, red, maroon, blue and green. Chou also can portray crafty court officials, commonly wearing a Guanyi in red or blue.

When Chou travel as officials, they may wear the Magua, usually in black, but sometimes in yellow or orange, with the Jianyi, often made of red fabric. Chou warriors are of lower rank than that necessary to wear the kao. Instead, they appear in versions of the Bingyi (soldiers' clothing)or Huakuaiyi (flowered fast clothes) in bright or dark colors depending on the nature of the character. A Chou rarely wears a Pi, but if he is playing an emperor or royal relative, a possibility when the character is foolish or corrupt, the Chou may wear a yellow Pi. The Chou also wears a red Pi when he marries. Chou of status often appear in the Huaxuezi. The most common color of the Xuezi they

---

❶ BONDS ALEXANDRA B. Beijing Opera Costumes: the Visual Communication of Character and Culture [M]. Honolulu: University of Hawaii Press, 2008: 83.

wear is a green that diverges from the green worn by honest and loyal characters. Chou perform servant roles dressed in plain black, blue or brown garments.❶

拙译 京剧中的丑角很少需要借助服装来体现人物的个性或地位的高贵。在少数情况下丑角会穿用蟒袍，一般是扮演太监的角色时，蟒袍颜色为绿色或橄榄绿，有时候也穿黄色、红色、栗色、蓝色和绿色的太监衣（蟒袍）。丑角还可以饰演狡猾的朝廷官员，通常穿着红色或蓝色的官衣。

当丑角以官员的身份出行时，他们可能会穿着黑色的马褂，但有时会是黄色或橙色的，且一般搭配红色的箭衣。扮演士兵的丑角因其级别卑微，不足以穿靠，而是根据人物角色的个性，穿颜色明快或暗沉的兵衣（士兵服装）或花快衣（绣有图案的快衣）出现。丑角很少穿帔，但是如果扮演的是皇帝或皇室亲戚，特别是当人物角色愚蠢或腐败时，就可能穿黄色的帔。当丑角结婚时，也穿红色的帔。丑角的社会地位通常借助花褶子来体现，最常见的颜色是绿色，但与诚实和忠诚的人物所穿的绿色不同。丑角在扮演仆人角色时，一般都身着纯黑色、蓝色或棕色的服装。

原文 The lower five colors are used in some of the official garments listed above, as well as in informal or domestic clothing. While two of the lower colors appear in court regularly, the others are more for informal wear and, as such, carry less significant meanings.❷

拙译 上面所列出的一些官服中的下五色，和家居或非正式场合穿着的衣服用色一样。下五色中的这两种颜色经常一起出现在宫廷中，其他下五色则更多见于非正式场合穿戴，并不包含太多深刻的蕴意。

（四）京剧服饰图纹文化内涵

京剧服饰图纹大致可以分为三类：神话题材、动物题材与植物题材。在神话题材中，龙纹作为最常见的纹样，多借鉴于明、清两代皇帝的龙袍。明黄色团龙图案的龙袍应用于戏衣之后，经过演变最终定型为艺术化的龙饰图案，成为权力的象征，并配以海水江崖与旭日图案，先是由皇帝专用，而后又泛化为帝王、将相通用，龙爪也统一为五爪。凤纹则象征吉祥、喜庆和爱情。凡是皇室女眷的服装，都用凤作为主要装饰图案。麒麟作为图腾崇拜，代表等级标识，象征位高权重。在动物题材中，蝙蝠纹常用于长者的服饰，蝠与福、富谐音，寄托了美好的祈福与祝愿，成为中国传统的吉祥纹样。在植物题材的图纹中，经常能看到"花中四君子"：梅、兰、竹、菊。这四种植物具有不畏严寒、刚正不阿的品格，文人士大夫常以此四君子借喻自己清高的气节

---

❶ BONDS ALEXANDRA B. Beijing Opera Costumes: the Visual Communication of Character and Culture［M］. Honolulu: University of Hawaii Press，2008：83–84.

❷ 同❶77.

和脱俗的情趣。笔者选取美国作者亚历山德拉关于京剧服饰图纹中类似动物或植物的英文描述，采用英汉对照的方法，挖掘服饰文化内涵，方便读者理解京剧服饰图纹的深刻寓意。

原文 Consolidated Patterns. Consolidated designs have four different manifestations, with arrangements of circles, single squares, yokes and bands. The circular motifs occur in several forms on traditional Jingju costumes. Considered the most perfect shape, the circle was originally reserved for the imperial family, but now it appears on the costumes of other higher-ranking characters as well. The content as well as the shape determines the status of the wearer. Round dragon motifs are worn by the highest-ranking Sheng (standard male), including the emperor and high court officials.

Round phoenixes are embroidered on the garments of some of the royal women. As early as the Tang dynasty, round flowered patterns appeared on garments. In traditional Jingju, similar round flower patterns are worn by Laosheng (mature men) and married women, either Qingyi (young or middle-aged women) or Laodan (mature women), in civil scenes. The prototypical round pattern repeats eight circles embroidered in a mixture of geometric patterns and flowers, dragons or phoenixes.

The round designs are placed on the robe with one at the center front and back at chest height, two at the knees in the front and back, and two on the sleeves. Square motifs are suitable for civil and military officers, so the square-shaped Buzi (rank badge) appear on the front and back of the Guanyi.[1]

拙译 依照形状布局，京剧服饰中的图案可分为四种，圆形图案（团花纹样）、单个方形图案（多为补子纹样）、上窄下宽型图案（常见于改良蟒）和条带形图案（边缘纹样）。团花纹在传统的京剧服装中有多种表现形式，被认为是最完美的形状。团花纹样（尤其是团龙纹样）原本仅为皇室留用，但现在它也出现在其他高阶层人物的服装上。图案的内容和形状决定了穿戴者的身份，团龙纹样一般由包括皇帝和高级官员在内的最高级别的生（男性）穿着。

团凤纹样通常出现在一些皇家女眷的衣衫上。早在唐代，服饰上就出现了团花纹。在传统京剧的民间场景中，类似的团花图案由老生（年长男子）、已婚女性，或者是青衣（年轻女子或中年女子）和老旦（年长女子）穿用。典型的团花图案一般8个为一组，由几何图形、花、龙和凤的刺绣装饰组合而成。

通常，团花分别出现在袍服的前胸、后胸的中心位置，以及前、后膝盖处，还有两处在袖口。方形图案适合文武官员，因此方形的补子（官衔的标志）出现在官衣的前胸和后背处。

---

❶ BONDS ALEXANDRA B. Beijing Opera Costumes: the Visual Communication of Character and Culture [M]. Honolulu: University of Hawaii Press, 2008: 98—99.

原文 Yokes in Qing garments are identified as areas embroidered in the shape of a yoke on a garment without an actual yoke seam. Yokes are a less common decoration in Jingju, but do occur in a few instances. For example, the black Mang often worn by Judge Bao frequently has a ring of circular ornaments stitched around the shoulders and tops of the arms in a yoke shape.

This use of the yoke pattern comes from the Qiu school of Jing performers, founded in the midtwentieth century by Qiu Shengrong (1915-1971), and the costumes for all of their official characters wear this pattern of embroidery. For women's dress, such a shoulder emphasis is accomplished with the addition of a cloud collar, an elaborately embroidered caplets with a standing band collar that is worn with the Nümang, Nükao (female armor) and Gongzhuang. This collar, with its curved edges and knotted fringe, is an item of Han dress traditions that continued into the twentieth century. The consolidated category also includes bands of embroidery. Embroidered bands appear on the women's skirt at the hem and around the front and back panels. The Nüxuezi (women's informal robe) with the center-front closing also may have embroidered bands down the center-front opening, around the hem, and up the side slits and the hem of the sleeves.❶

拙译 清代官服的叉形部分并没有实际的裁片缝线，仅是刺绣的框架式样。在京剧中叉形装饰并不常见，仅在几场合出现。如在包拯穿的黑色蟒袍中，有时会用圆形的装饰缝在肩膀四周，在手臂的上部也有一个叉形的装饰。

对于叉形图案的使用源自20世纪中期京剧裘派表演艺术家裘盛戎（1915—1971），他塑造的所有官员人物都穿用这样图案的服饰。女性服饰如果也想要强调肩部，则通过添加云肩，将精巧的立领刺绣坎肩和女蟒、女靠以及宫装搭配在一起。这种领子的边缘处有弧度，配有结，是汉朝传统装束在20世纪的延续。其综合类的使用还包括装饰的带子，经常出现在女性裙装的边缘以及前后片。如褶子（女性的非正式装束）前片的中缝处装饰有刺绣的带子，一直顺垂到前片中间的开口处，还有褶子的底边、摆衩以及袖口边缘。

原文 Integrated compositions combine motifs over the entire surface of the garment in a unified theme. Where Qing court garments use celestial, terrestrial and panoramic vistas for this category, traditional Jingju costumes draw primarily from the terrestrial motif used on the historical garment known as a dragon robe (Longpao, later renamed Jifu). This garment was viewed as a representation of the universe, and its distinctive embroidery pattern has been transferred to the Mang for stage usage in court scenes. The components of terrestrial composition that the imperial and theatrical garments share include the hem areas decorated with parallel diagonal straight or

❶ BONDS ALEXANDRA B. Beijing Opera Costumes: the Visual Communication of Character and Culture [M]. Honolulu: University of Hawaii Press, 2008: 99-100.

wavy lines representing standing water (Lishui), topped by concentric semicircles representing still water (Woshui). Emerging from the water, at the four axes of the robes, are trapezoidal mountains representing the four cardinal points on the compass.❶

拙译 综合运用不同图案，并将整个服装表面的图案统一为同一主题。清朝宫廷服的主题是利用天、地以及全景，传统京剧服装图案也主要取材于地的主题，运用于龙袍后改名为吉服）这样的历史服装上。服饰被看作是宇宙的代表，典型的刺绣图案被演绎为蟒，并在舞台的宫廷剧中运用。帝王袍和戏剧中的服饰纹样类似，下摆处都是用平行的条状斜纹或波浪线代表海水（立水），海水上方集中的半圆代表着海浪（卧水）。在长袍的四条轴线上，露出的是代表四个不同方位的江崖。

原文 Originally, the mountains were dominant, but gradually the depth of the waves increased to cover the lower part of the robe to the knees, and the mountains decreased in size. Dragons and clouds decorate the remainder of the surface to complete the terrestrial theme.❷

拙译 最初，袍服上的江崖图案占主体，但之后海水图案逐渐加高，直至袍服的膝盖处，山形变小。在江崖海水纹之上装饰有龙纹和云纹，彰显袍服的主题图案。

原文 Two significant differences in the embroidery occur between the historical and stage garments. In imperial dress, originally only the emperor could wear a garment with a five-clawed dragon (Long). It was not suitable for costumes to copy the emperor's symbol, so the theatrical version of this garment originally had a four-clawed dragon (Mang) on it. Such embroidery was also worn by nobles. The written character Mang is now the word for python, and the Jingju robe derives its name from this creature. Since China no longer has an emperor, the Mang robe often has a five-clawed dragon, but it is still called a Mang, and hence the translation to court robe rather than dragon robe.

The second distinction comes with the twelve symbols of imperial power that are scattered into the composition of the historical dragon robe. Because these symbols were also reserved exclusively for the emperor's use, the surface of the Mang in traditional Jingju employs the eight emblems of Daoism, Buddhism or Confucianism instead.❸

拙译 历史版龙袍与舞台版龙袍的最大区别在于，历史上最初只有皇帝可以穿五爪龙的帝王袍，但是京剧戏服效仿皇帝的象征显然不妥，因此最初戏服中的龙袍只能用四爪龙（蟒），且这样的刺绣图案仅限于贵族。蟒现在的字面意思是动物巨蟒，京剧

❶ BONDS ALEXANDRA B. Beijing Opera Costumes: the Visual Communication of Character and Culture [M]. Honolulu: University of Hawaii Press, 2008: 100.

❷ 同❶99–100.

❸ 同❶100–102.

中蟒袍的名字源于此。因为中国现在不再有皇帝了，蟒袍上也经常出现五爪龙的图案，但是依然被称为蟒，并被翻译为宫廷服，而非龙袍。

第二点区别来自龙袍上象征帝王权力的十二章纹，在传统京剧中，蟒袍上绣有来自道教、佛教或儒家的八种象征图案，以象征帝王身份。

原文 Flowers are a dominant image in traditional Jingju dress, and most of their arrangements on the fabric will include more space above the subject than below, as the unornamented area above links to heaven and that below to the earth, and the floral subject, being of the earth, connects earth to heaven. Two-thirds of the surface is generally left blank to focus on the essential simplicity of the image, and most of the subject of the picture is confined to the lower segment of a diagonally divided picture space. The pattern of placement from the paintings is reflected in many of the floral fragments on costumes. Asymmetry logically represents the essence of nature, being a more dynamic reflection of the natural growth of plants. While the individual motifs are asymmetrical, a design can be reversed and repeated on the two front panels of a Nüpi, creating an overall symmetry from asymmetrical designs.❶

拙译 花卉在传统京剧服饰图案中占据主导地位，面料上花卉图案的上下空间留白也有讲究，一般在花卉图案的上面位置，相较于下面的留白空间要更大一些。这是因为在中华传统理念中，花卉作为主体图案，其上面没有装饰的区域与天相连，下面的区域则与地相连。同时，通常面料表面三分之二的部分都留白，以保持图案整体的简洁性，而花卉大多定位在面料中心偏下的位置，通过花叶和花朵这些细微之处展示服装上花卉的布局模式。比如，不对称的特点就体现了自然的本质，即花卉在自然生长中不断变化的天性等。虽然每朵花都是不对称的，但设计图案时可以在女帔的两个前片上颠倒和重复，从不对称的设计中创造出整体的对称。

原文 The area of the field or the garment piece, determines the size and number of the flowers. Odd numbers of blooms are considered more auspicious than even numbers, and flowers in different stages of growth are incorporated to add to the natural effect. A single curving stalk of five flowers 3 to 4 inches in diameter can be arranged into a balanced composition on the back of a Nüpi. On the Xuezi, the surface provides a larger area for decoration and voids. For young scholars, the flower fragments are placed on their Huaxuezi on the left side of the hem and the right side of the chest, around the asymmetrical closing.❷

拙译 京剧服饰上的图纹犹如苗圃中的花卉，田地的大小或衣服的大小决定了花

---

❶ BONDS ALEXANDRA B. Beijing Opera Costumes: the Visual Communication of Character and Culture [M]. Honolulu: University of Hawaii Press, 2008: 102.

❷ 同❶102-103.

京剧服饰文化传播及其文化内涵

第二章

41

朵对应的尺寸和数量。但是通常认为花朵数量为奇数比偶数更为吉祥，并且由于花朵处于不同的生长阶段，受自然影响，数量会不断增加。由五朵直径为7~10厘米的花组成的弯曲茎排列在一起，就可以构成女帔背面的平衡组合图案。在褶子表面，似乎有着更大的装饰空间。对于年轻书生角色所穿的花褶子而言，花瓣的布局大多位于服装下摆的左侧和胸部的右侧，呈不对称闭合状。

原文 Traditional Jingju costumes absorbed some of the symbolic meanings connected with the subject matter of the images. Though the costumes are not meticulously recreated from information in imperial edicts concerning dress, many of the basic concepts of cultural symbolism are still honored. For example, civilian costumes generally have floral and bird imagery, while animals are used for military men, as was the case in history.❶

拙译 传统的京剧服饰借助与图案主题相关的象征意义，彰显文化内涵，尽管这些概念与内涵并非"君权神授"或"绝对真理"，但依然属于大众认同的范围。例如，平民服饰通常绣有花卉和鸟类图案，武士的服饰则通常绣有动物图案，历史上也确实如此。

原文 The conventions of dress are mutable within role type, but not across role categories. As theatrical garments can be used for a variety of characters within the role type, the garments are embroidered with generic flowers or rank symbols suitable for the range of men and women who may wear them. In addition, while most of the imagery comes from nature, the objects are idealized and stylized.❷

拙译 京剧服饰装束在同一角色类别中并非一成不变，戏剧服装因其可适用于不同类型角色，故服装上常绣有适合不同性别且可通用的花卉或等级符号。另外，尽管大部分图案取于自然，但个体表达的呈现却属于理想化和程式化。

原文 Lions and tigers are depicted in a codified style and appear alongside of fanciful composites, such as the dragon and phoenix. Just as some of the animals are invented, the flowers are subject to the artist's creativity as well. Some flowers are recognizable blooms, while others are generic.❸

拙译 狮子和老虎常以系列风格呈现，并与幻想中的动物并列，如龙和凤凰。犹如动物被创造一样，这些花卉也受到艺术家创造力的影响。有些花卉属于大家熟知的知名花卉，而另一些则较为普通。

---

❶ BONDS ALEXANDRA B. Beijing Opera Costumes: the Visual Communication of Character and Culture ［M］. Honolulu: University of Hawaii Press，2008：103.

❷ 同❶103-104.

❸ 同❶104.

原文 Traditional Jingju costumes primarily utilize four of these categories: floral, faunal, imperial and emblematic. Floral images indicate the harmonious existence between humans and nature, and often represent wishes of abundance and success in many areas of life. Faunal images include both animals and birds, which often carry symbolic connotations. Imperial imagery appears on court dress and includes a dragon, placed within a cosmic landscape. The emblematic category contains three groups of eight meaningful objects related to Daoism, Buddhism and Confucianism. Their placement on a garment evokes social success or good fortune. Emblematic images also include the rebus, a distinctive punning device unique to the Chinese language.[❶]

拙译 传统京剧服饰主要利用花卉、动物、帝王和象征这四类图案表达内涵。花卉图案表明人与自然之间的和谐存在，祈愿人们的生活富足成功。动物图案包括动物和鸟类，通常具有象征意义。帝王图案包括宇宙景观中的龙，一般出现在朝服上。象征类的图案分为三组，包含八个与道教、佛教和儒学有关的内涵丰富的物件，给人以有关幸福和成功的联想。同时也包括字谜或画谜，属于汉语中双关语的独特表达。

原文 Floral Imagery. In Chinese culture, women's informal garments were commonly embroidered, where, apart from dragon robes, Chinese men rarely wore embroidered clothes. Roundel patterns, butterflies, floral and fruiting sprays, and garden vignettes were reserved for women's dress.

By contrast, in the traditions of Jingju costumes, men in several role types frequently wear garments with embroidered images. Floral designs are most likely to occur on informal dress, including the garments worn by many of the characters at home and at noncourt functions, the Xuezi, Pi and skirt.[❶]

拙译 花卉意象。在中国文化中，女性的非正式服装通常带有绣花，除了龙袍之外，中国男性很少穿绣花衣服。团花、蝴蝶、开花结果形，以及园林景观都是女性专属图案。

与此形成鲜明对比的是，在传统京剧服饰中，不同角色类型的男性也经常穿着带有刺绣图案的服装。非正式礼服中最有可能出现花卉图案，包括在家和不需要上朝时穿用的服装，即在褶子、帔和裙子之类的服饰上也有花卉图案。

原文 In the case of traditional Jingju costumes, the brighter colors of thread are needed in the embroidery not only to contrast with the some of the intense colors of the garments, but also to project the designs to the audience. The vivid colors are also joined in unique combinations on

---

❶ BONDS ALEXANDRA B. Beijing Opera Costumes: the Visual Communication of Character and Culture [M]. Honolulu: University of Hawaii Press，2008：104.

individual garments. A single stalk may contain flowers of different colors, with as many as three hues being quite common. A specific flower can also be rendered in a variety of colors that are not limited by the natural colors of that flower. The most popular colors for flowers seem to be reds, including peach, cerise, and pink, as well as purple, blue, and yellow. Leaves tend to be in the blue-green range, although some examples have a more leaf-green color. The graceful, elegant overall effect of the stylization and color of the flowers in informal dress creates a sense of tranquility for the domestic scenes. In addition to the artistic elements of the decorations, many flowers contain symbolic and auspicious messages. Flowers represent the four seasons: the plum blossom for winter, peonies or orchids for spring, the lotus for summer, and the chrysanthemum for autumn.❶

拙译 在传统京剧服饰中，通过彩线刺绣不仅可以形成强烈的色彩对比，还能将设计鲜明地呈现出来。有时亮丽色彩的图案在个体服饰上也会形成独特的组合。如单根茎上可能出现不同颜色的花朵，三种颜色以上的花卉也相当常见，其中红色，包括桃色、樱桃色、粉色，以及紫色、蓝色和黄色似乎最受欢迎。特定的花卉有时甚至不受自然花色的限制，尽管花叶一般为绿色，但是刺绣中的叶子更趋于蓝与绿之间的过渡色。非正式礼服中刺绣花卉的程式化，以及色彩的优雅，为京剧中家庭场景的创设增添了特有的宁静。花卉作为艺术元素，还具有吉祥与象征意义。如鲜花用来代表四季：冬天的梅花，春天的牡丹和兰花，夏天的莲花，以及秋天的菊花。

原文 As there are no seasons indicated in traditional Jingju, the seasonal meanings of the flowers are not used, but each flower conveys other messages, as well. Plum blossoms stand for resistance and the will to live because they bloom in the winter. Peonies, with their bold beauty and commanding size, carry meanings of wealth and advancement. They are variously assigned meanings of brightness and masculinity, love and feminine beauty, and as an omen of good fortune.

The lotus, in a concept that comes from Buddhism, represents purity and nobility, as its white blossoms emerge unsullied from the mud. The chrysanthemum appears quite often, being valued for its variety and richness of color. It is regarded as the national flower and the symbol of noble personalities, gentility, fellowship and longevity.❶

拙译 传统京剧中四季不分，因此花卉的使用不是表达季节，而是用来传递一些其他信息。冬季凌霜绽放的梅花象征反抗和生命力；娇艳傲骨的牡丹则代表富贵和升迁。不同花卉具有不同的寓意，诸如聪明和刚毅、爱情和柔美，以及相伴的好运等。

源于佛教理念的莲花，出淤泥而不染，代表纯洁和高贵，常被用于衣饰图案。并因其形态各异、色彩丰富而备受重视，成为佛教诞生地的国花，常被用来象征高贵的人格、文雅的气质，以及友谊和长寿。

❶ BONDS ALEXANDRA B. Beijing Opera Costumes: the Visual Communication of Character and Culture［M］. Honolulu: University of Hawaii Press, 2008: 106.

原文 In addition to these four, roses,which appear regularly on young women's robes, are a symbol of eternal youth and lasting springtime. Principal characters sometimes have characteristics specifically communicated through the flowers on their robes, but because the costumes are interchanged among characters within a role type, for most characters the flowers are merely ornamentation indicating gentility.[1]

拙译 除了以上四种花卉以外，在年轻女性的长袍上还经常可以看到象征青春和春天的玫瑰图案。京剧戏曲中的主角有时需要借助身着绣有花卉图案的长袍来表达其性格特点。但是由于在京剧中，同类角色彼此之间的服装可以交换穿用，所以对于大多数角色而言，花卉图案仅是角色人物高贵出身的象征。

原文 Faunal Imagery. In Chinese culture, specific animals carried desired qualities and were embroidered onto items of clothing to endow the wearer with those attributes. For example, the tiger, embodying courage and fierceness, appears on the garments of a warrior. In other cases, the creatures convey wishes for a good life, as with the crane, which has come to symbolize longevity.

Symbols for behavioral characteristics are assigned specifically to either men or women, but animals representing social wishes can appear on the dress of either gender. The faunal language on women's clothes is less complicated than that on men's garments because floral imagery predominates on women's garments.[1]

拙译 动物意象。在中国文化中，不同的动物都有其各自不同的特性，作为图案刺绣出现在衣物上，也赋予了穿着者这些属性。例如，代表着勇敢和凶猛的老虎常出现在武士的服装上。而其他动物，则传递出了着装者对于美好生活的期望，如鹤象征着长寿。

虽然作为行为特性的符号对于男女服饰具有不同意义，但是这种传递美好期盼的动物形象图案都可以出现在不同性别的衣饰上。同时，因为花卉意象在女性服装中占据主导地位，所以女性服饰上的动物意象通常较男性更为简单。

原文 Animal imagery appears on the male garments including the Guanyi, Bufu (coat with a badge), Kao and Kaichang (informal official robe). The Guanyi resembles the garment of rank developed from the Ming dynasty tradition and is now worn by traditional Jingju officials of the court. A long robe with an asymmetrical closing, the historical garment was relatively free from ornament except for the Buzi that were placed on the center of the chest and back. The rulers from the Mongol tribes of the Yuan dynasty (1271-1368) had worn square decorations woven into their

---

[1] BONDS ALEXANDRA B. Beijing Opera Costumes: the Visual Communication of Character and Culture [M]. Honolulu: University of Hawaii Press, 2008: 106.

robes, although these ornaments were not associated with rank, but the Buzi were instituted as insignia badges in 1391, in the Ming dynasty.❶

拙译　动物意象出现在包括官衣、补服（带补子的外套）、靠和开氅（官方非正式长袍）在内的男性服装上。官衣从明朝发展而来，类似于当时象征等级的服装，为传统京剧官员在朝堂上穿用。不对称且闭合的长袍，除了在胸部和背部的中央配有补子之外，几乎没有其他装饰性元素。元朝（1271—1368）时蒙古部落的统治者曾将方形装饰织入袍中，虽然这些饰物与阶级并不相关，但在1391年的明代，这种方形的装饰图案被设立为代表官职的补子。

原文　As the circular shape, often manifested in dragons and phoenixes, was employed primarily for the royal family, the square form of these badges became emblematic for the court officers. While the right to wear a rank badge was bestowed by the emperor, recipients had to provide the badge themselves, and hence a tradition developed of having the badges made separately from the garment and stitched on. Since the officers were likely to have aspirations for a higher rank, the symbol indicating the rank was often embroidered on a separate, interchangeable piece as well.❷

拙译　由于圆形补子通常为龙和凤的图案，并主要用于皇室，所以这些方形的补子就成了朝廷官员的标志。虽然只有皇帝才有授予官衔的权力，但接受官衔者则要自己制作补子，因此出现了将衣服与补子分开制作与缝合的习俗。由于官员通常都渴望晋升，所以作为官衔标志的补子经常处于独立位置，以方便衣服的刺绣部分因官位变化而更换。

原文　There were two categories of rank: civil for scholars and officers of the court, and military for men of action. Within each of these two categories, there were nine degrees of official rank, the first being the highest. The garment for each rank was delineated both by the color of the robe and by the subject in the square of embroidery on the front and back of the garment.

Birds, symbols of elevated intellect, were used to distinguish the civil ranks, and animals, indicative of physical strength, were employed for the military ranks. It is not known how the creatures were selected to represent their specific ranks, other than those connections that may be found in Chinese mythology.❸

拙译　京剧服饰中的官员等级分为两类：学者或文职官员，以及武将。无论文职还是武将，都分为九个官阶，一等为最高级别。每个等级的官服都是通过袍服的颜色

❶ BONDS ALEXANDRA B．Beijing Opera Costumes：the Visual Communication of Character and Culture ［M］．Honolulu：University of Hawaii Press，2008：106．

❷ 同❶106–107．

❸ 同❶107．

以及服装正背面刺绣图案的纹样来体现。

　　作为智慧象征的鸟类，在服饰图案中用来区分文官的官阶；而作为力量象征的兽类，则用来区分武将的官阶。至于为什么这些动物能代表官员的特定官阶，无从而知，也许只有中国神话能带来些许启迪。

　　原文　The birds and animals were placed in a landscape facing towards the sun, an homage representing the emperor. The sun was located in different corners for civil and military officials: the birds faced the wearer's right, and animals faced the wearer's left. The square patch was outlined with a frame to set it off. In the court, the display of the proper rank on one's garments was quite important, but for traditional Jingju usage, precise rank no longer carries significance. Generally, only civil officers wear the Guanyi onstage, so the badges are primarily embroidered with birds, although some Guanyi have badges with a sun in the center for a more generic emblem. The Bufu is the Qingdynasty version of the garment with a rank badge, and it is used onstage for officials of foreign courts.❶

　　拙译　京剧服饰中鸟和兽的图案都要放置于面向太阳的位置，以表达对皇帝的敬意。但是同样一个太阳，在文官和武将的官服中的位置也略有不同，文职官服上的鸟类一般头朝穿衣者右侧的太阳，而武将官服上的兽类则一般头朝穿衣者左侧的太阳。这些鸟类或兽类的图案都被放置在官服前后片上正方形的补子之中。在朝堂上，官服的等级标志极其重要，但是对于传统京剧而言，精准的等级不再具有重要意义。通常，只有文职官员在台上身着官衣，尽管一些官衣会以太阳为中心来表达特定的象征意义，但是补子主要局限于鸟类。在清朝，补服是官服等级的象征，所以在京剧舞台上成为中国官员的标志。

　　原文　The Kao developed for use as theatrical armor and the composition of the designs on the surface is unique, creating one of the most highly ornamented garments in the wardrobe. With elaborate embroidery on both the male and female versions, the animal and bird imagery projects the power of the wearer. Dragons are depicted on Kao of the emperor's warriors. Symmetrical paired dragons can be featured on the padded front belt piece (Kaodu), or a single dragon may face sideways or appear with its head in the center of the band.❶

　　拙译　靠是从京剧服饰中的铠甲发展而来的，在戏服设计中独一无二且最具装饰性。靠作为装饰，在男女戏服中通用，鸟和兽的不同图案暗示了穿着者的不同权力。皇帝身边的武士身着带有龙的刺绣图案的靠，对称的双龙可以绣在前面的靠垫（靠肚）上，单龙可能面向侧面或其头部位于靠肚中心。

❶ BONDS ALEXANDRA B. Beijing Opera Costumes: the Visual Communication of Character and Culture［M］. Honolulu: University of Hawaii Press, 2008: 107.

京剧服饰文化传播及其文化内涵

第二章

原文 Dragons can also be used to embellish the hanging panels and the lower sleeves. In addition, the torso pieces, the upper sleeves and the some of the panels below the waist are often decorated with a scale pattern that evolved from the use of segmented plates to facilitate movement in historical armor. A geometric design may be used in this area as well.❶

拙译 龙也可以用来装饰京剧服饰中垂挂的部分和袖子下部。此外，躯干部分、袖子上部和腰部以下的一些部位通常会装饰鳞片图案，该图案用分段片悬吊的方式呈现出来，方便古代士兵的活动。同时，几何设计在此得到运用。

原文 Tigers are frequently used as motifs on the armor of characters who are uninhibited and straightforward. Large dimensional tiger heads can fill the space on the padded belt, and some forms of Kao have dimensional tiger heads on the shoulders as well. A pair of fish tails finishes the lowest point on the apron panel over the legs in the front of the costume.The two tails evolved from the Yin/Yang symbol of two fish swimming in a circle. The fish emblem also means that the character thinks of clever strategies. An embroidered border emphasizes the shapes of all the panels. The border may contain ancient-style smooth dragons, geometric patterns of water, or a meander resembling the Greek key, called cloud-and-thunder by the Chinese.❶

拙译 在京剧戏服中，老虎常作为放荡不羁与性格直率的角色象征而用于铠甲的图纹上。大尺寸的虎头可以填充在带衬垫的腰带空间里，而靠的肩头部位也常绣有立体的虎头。靠的下甲最低处为一对鱼尾形状，成对的鱼尾是从阴阳太极图，即两条形成环型的正在戏水的鱼演变而来，因此鱼的图案也成为有谋略的戏曲人物的标志。刺绣图案的边线常被用来强调图纹形状，如可以代表古代飞舞的龙。此外，几何形代表水，类似的蜿蜒的希腊花边在中国人眼里又是云与雷的象征。

原文 For female versions of armor, the phoenix, rather than the dragon, is the primary creature used, as women's armor is decorated with flying creatures, rather than the earthbound animals used for men's. The phoenix appears in the same locations as the dragon on the male armor, as well as on the larger cloud collar. Flowers, often peonies, ornament the additional narrow streamers and the borders on the edges of the shaped pieces. The embroidered designs on the rest of women's armor are similar to those on men's, with scales in the central parts of the panels and borders defining the edges. Women's armor generally has a greater variety of colors in the embroidery than does the men's, which often is dominated by metallic couching to make it look impermeable. The Kaichang is worn by military and official men in their private quarters. The

❶ BONDS ALEXANDRA B. Beijing Opera Costumes: the Visual Communication of Character and Culture [M]. Honolulu: University of Hawaii Press, 2008: 107.

Kaichang for Jing features large, bold animal figures often stretching the length of the garment.[1]

拙译 京剧戏服中女性的铠甲，主要是选用凤、而不是龙来作为装饰的。女性的装饰图案一般是飞禽，而男性的则更多是走兽。凤在女性铠甲中的位置类似于男性铠甲中龙的位置，都出现在较大的护领（苫肩）上。花朵，通常是牡丹，装饰在另加的窄飘带与固定物件的边缘处。女装铠甲中其余部分的绣花图案与男式铠甲上的刺绣图案相似，中心位置有鳞片，而且边缘处界限分明。但女性铠甲上的刺绣通常比男式的色彩更加丰富，而且多用金属色，使其看起来更致密。开氅是武士或武官在非正式场合的装束，京剧中的开氅常利用象征勇猛的动物图案来延展袍服的视觉效果。

原文 Animals pictured include lions, leopards, tigers, elephants and the Qilin. The Qilin is a composite creature, often called a dragon horse or the Chinese unicorn, as it is sometimes depicted with a single horn emerging from its forehead, though some representations have two horns. The animal parts that comprise the Qilin include the body of a deer, the tail of an ox, and the hooves of a horse. The body has scales with a ridge down the back, and the head has a flowing mane and two trailing antennae. The early Qilin was an aggressive beast with the capacity to discern good and evil. Although the qilin later became more benevolent and not harmful to living creatures, it represented the first grade of military officers in the Qing court. In traditional Jingju, military men rarely wear their rank badges, but Qilin imagery appears on their Kaichang.[2]

拙译 袍服上的动物图案包括狮子、豹子、老虎、大象和麒麟。在中国古代传说中，麒麟与龙、凤、龟合为四灵，乃毛类动物之王。通常被称为龙马或中国麒麟，尽管它是两角兽，但是有时也被描绘为前额一个角。麒麟由鹿的身体、牛的尾巴和马的蹄子组成。脊背有鳞片，头部有流鬃和两条触须。早期的麒麟具有辨别善恶的能力，并具有侵略性，后来变得仁慈而不伤人畜，便用来代表清朝的一等武官。因此，在传统的京剧中，武官很少佩戴他们的徽章，但麒麟图案常在他们的开氅上出现。

原文 Imperial imagery. Probably inspired by an alligator, the dragon is associated with water and rain, and therefore with ancient emperors, who through performing rituals properly, were responsible for ensuring enough water and rain to guarantee crops to feed their subjects. The first reliable reference to dragons as the principal design on these robes dates from the Tang dynasty. Dragon designs come in many shapes and sizes, the dominant arrangement of the dragon's body being a serpentine twist to form a roundel, repeated several times over the surface of the garment. On garments with these round dragons, there are usually ten roundels dispersed over the surface

[1] BONDS ALEXANDRA B. Beijing Opera Costumes: the Visual Communication of Character and Culture [M]. Honolulu: University of Hawaii Press, 2008: 107-108.

[2] 同[1]108.

of the mang, one at the center of the chest and back, two above the wave patterns on the front and back and two on each sleeve. This placement pattern comes from Han design, rather than the Qing, which used only nine dragons. The central dragons at the chest and back are generally arranged with the head facing forward at the top of the circle; the dragons near the knees are in profile facing each other. Gentle, intelligent, and courageous male characters and young generals who are handsome and elegant wear the mang with rounded dragons.[1]

拙译　龙是帝王的意象，或许是受鳄鱼的启发，龙又与水和雨有关，因此古代皇帝常通过例行的仪式，祈祷确保有足够的雨水保证万物生长，以养活黎民百姓。龙作为长袍设计的主要图案首先来自于唐代。龙的设计有多种形状和大小，龙身的主要部位呈蛇形扭曲，形成圆圈，在服装表面重复数次。在这些团龙图案的服装上，通常有十个圆形图纹分布在蟒袍的表面，在胸部和背部中央各一个，在正面和背面的波浪图案之上各两个，每个袖子分别有两条龙。这种龙纹的布局模式来自汉族的设计，并非来自只用九条龙的清朝。胸部和背部中央位置的龙，通常其头部朝向前方排列，靠近膝盖的龙则彼此面对。一般儒雅、聪明、勇敢的男性角色，以及帅气、优雅的年轻将军穿着带团龙的袍服。

由上可见，京剧服饰图案内涵丰富。首先，京剧服饰图案创造的意象画面反映了现实生活，体现的是"天人合一"。其次，京剧服饰图案具有超越现实的特征，如京剧中富贵衣的补丁图案，剧中的丝绸面料与彩色补丁并不意味着生活中的贫穷。这种穷衣，图有穷之"意"，而无穷之"实"。同时，京剧服饰图案在创作过程中充分体现出真、善、美三者之间的和谐统一，在符合人物身份与性格特点方面求真；在寄托美好祝愿和体现人物内心世界方面求善；在展示舞台效果和尊重戏剧人物、戏剧作者和观众等诸多矛盾统一的过程中求美；真、善、美的和谐统一贯穿于京剧服饰的始终。

本章小结　｜　京剧形成于清代，随着剧情发展的需要，其服饰基于唐、宋时期代表性服饰的特点，同时吸纳满族的服饰风格，最终形成以明代服饰款式为主的京剧服饰——行头，同一套行头在不同的剧目里，表现不同的角色，映射出剧中人物的性格。京剧行头通常可以在不同的情形下穿着，无时代、地域或季节之限制。因此，京剧服饰研究对于理解京剧剧情，加强传统京剧文化的对外传播意义重大。本章聚焦中外京剧服饰文化研究现状，阐释京剧衣箱分类，关注传统京剧服饰特点与文化内涵，包括京剧服饰颜色和图纹，京剧中不同人物角色与服饰色彩，针对京剧传播现有的多种中英文版本归类梳理，逐一对照呈现，尝试借助京剧服饰文化协调翻译原则，补充完善当今文化语境下的传统京剧服饰文化内涵，体现京剧服饰文化在"天人合一"中呈现出的"真、善、美"和谐统一。

---

[1] BONDS ALEXANDRA B. Beijing Opera Costumes: the Visual Communication of Character and Culture [M]. Honolulu: University of Hawaii Press，2008：108.

第三章

京剧服饰大衣箱文化阐释及翻译协调美

第一节

京剧服饰主要形制概述

京剧服饰的主要形制可以分为三大类，主要包括大衣、二衣和三衣。大衣是文职官员、老爷太太、小姐丫鬟、书生等角色的服饰总称，是京剧服装内部分工的行当之一，在技术职能方面，主要负责管、扮、扎、勒，即服装的管理保养，同时在演出中负责演员的服饰装扮以及特殊人物的扎勒；大衣在管理上有很强的技能操作，对于服饰名称识别和塑造不同人物着装类型而进行的服装分配起到了保证作用。二衣是元帅将官、英雄豪杰、绿林好汉、壮丁打手等角色的服饰总称，二衣的范围根据大衣行当所分工的范围而形成，它们之间在技艺处理上虽有共同之处，但在其他方面差异明显。三衣是剧中人物所穿的内衣及靴鞋等物的总称，根据服饰的用途、用料、表演行当进行分类，如旦角服装、老旦服装，或丝绸的、软缎的等。通过长期实践，京剧服饰形成了一套较规范的管理体制。下面笔者摘录了部分潘霞凤著作中有关京剧服饰的描述，并采用英汉对照的方式呈现如下。

原文 A costume wardrobe (Quanxiang) in Peking Opera generally implies a whole set of clothes and ornaments performers are to wear in a show, with a variety of containers called in Chinese Xiang or Xia for both theatrical apparatus. They may be arranged in the following sequence: The primary costume trunk (Dayixiang) holds all the principle articles (Dayi) players have to put on for a performance, such as the ceremonial and informal robes, the official robe, the informal gown (Kaichang), the commoner's coat, etc., which are characterized by a pair of cascading sleeves. Boxes which are part of this range fall into two sets, namely, the upper and the lower, and their Chinese appellations are Shangshouxiang and Xiashouxiang. While the upper set is exclusively for the VIP's attire, the lower one contains mainly the commoners' dress, including the scholar's garment (Xueshiyi), the eight-diagram garment (Baguayi), the Taoist nun's garment (Sengyi) and the like, as well as the above-mentioned patched garment and commoner's coat.[1]

---

[1] PAN XIAFENG. The Stagecraft of Peking Opera [M]. Beijing: New World Press, 1995: 110.

52

拙译 京剧服饰一般指舞台表演的全套服饰，称全箱，包括剧院的各种箱或匣。一般依照下列次序排列：最重要的衣箱（大衣箱）装着表演时穿戴的最基本的服饰（大衣），如正式或非正式场合的袍服、官服、非正式的袍（开氅）和平民的外衣等，这些衣服共同的特征是有着层叠的袖子。这类衣箱分为两部分——上下两层，即上首箱和下首箱。箱子上层专门为贵宾服装而设，下层主要是百姓服饰，包括学士衣、八卦衣、僧衣等，以及上述的富贵衣和普通外衣。

原文 In the past, there were strict rules and regulations as to which of the articles should be put above another. Ironically, however, a royal robe usually worn by an emperor could in no way lord it over the whole set of articles and had to be relegated to the third place in the trunk so far as there were red robes worn by the Heavenly Officials (Tianguan) or God of promotion (Jiaguan), which should be put at the very top, and, next to them, a greenish one worn by Mammon, God of Wealth (Caishen). As a result, this tradition had led to the present disposition of robes in the upper trunk on the basis of the five primary colors (Shangwuse) red, green, yellow, white and black in order. The lower trunk is as a rule topped with a patched garment which is followed by a drab coat.❶

拙译 过去人们对于衣服放置的上下位置有着很严格的规定。然而，具有讽刺意味的是，通常皇族的袍子由皇帝穿用，但也没办法放在整个服饰的最上面，而是天官或加官时穿的红袍子放在最上面，紧接着的是财神穿的带绿色的服饰。到现在为止，龙袍也只能屈尊放在衣箱的第三层。这样的传统导致目前上箱衣物的放置遵循五种基本色彩（上五色）排序，即红、绿、黄、白和黑。下箱的规矩是上面放补服，下面放黄褐色大衣。

原文 Also made up of an upper and a lower sub-grouping, the secondary trunk (Eryixiang) holds form top to bottom such items as armor (Kao), the dragon-on-flower-decorated archer's dress (Longjian or Huajian), the hoer's suit of a jacket and a pair of pants (Kuayi and Kuaku), and a blouse and trousers which are regularly placed at the bottom. Articles pertaining to the secondary trunk will generally be labeled the acrobat's wear (Wufu).❷

拙译 二衣箱同样由上下层组成，从顶层到最底层放置的服饰有靠、团龙纹和团花纹装饰的箭衣（龙箭或花箭）、男士成套的衣裤（侉衣和侉裤），以及一般都放在箱底的罩衫和裤子等。属于二衣箱的服饰一般都标记为特技演员的衣服（武服）。

原文 What is typical of the tertiary trunk is that it is always a single one, holding such

---

❶ PAN XIAFENG. The Stagecraft of Peking Opera［M］. Beijing：New World Press，1995：110–111.

❷ 同❶111.

matching accessories as a sweat shirt (Shuiyizi), padded waistcoat (Pang'ao), colored pants (Caiku), collar guard (Huling) and so on.❶

拙译 三衣箱的特点是放置些单品，如汗衫（水衣子）、棉袄（胖袄）、彩裤、护领等配饰。

原文 In addition, there is a headdress trunk (Kuitouxiang) for storing items like a crown (Guan), helmet (Kui), turban (Jin), hat or cap (Mao) as well as various kinds of artificial beards; a flag and drapery trunk (Qibaoxiang) for such props as the larger curtain (Dazhangzi), the smaller curtain (Xiaozhangzi), the flying tiger flag (Feihuqi), the eight-diagram flag (Baguaqi), the tablecloth (Zhuowei), and the chair covering (Yipei); and footwear trunk (Xuexiang) for shoes and boots.❶

拙译 除此之外，有盔头箱，用于放置冠、盔、巾、帽以及其他各种人造髯口。还有用来放置器具和什物的旗包箱，这些道具有大帘子（大帐子）、小帘子（小帐子）、飞虎旗、八卦旗、桌帷和椅帔，以及放置鞋子和靴子的靴箱。

第二节
京剧大衣箱服装文化内涵及其协调翻译

大衣箱在京剧衣箱中的地位举足轻重，带有水袖的服饰大都属于大衣箱类，主要用于塑造宫廷帝后、朝廷权臣、地方官员、后宫妃嫔，以及书生雅士、达官显贵、老爷夫人、少爷小姐，乃至丫鬟仆人等人物的舞台形象。大衣类服饰主要包括男女蟒、官蟒、改良蟒、旗蟒、开氅、官衣、宫衣、男女帔、花素褶子、腰裙、袄裤、太监衣、僧衣、道衣、八卦衣、罪衣、罪裤和孝衣等五十余种。另有副大衣箱，用于放置人物装束搭配使用的装饰配件和服饰，主要包括斗篷、饭单、四喜带、丝绦、腰巾等，配饰之间可以互相搭配，共同塑造舞台人物形象。

依照梨园行规，大衣箱的最上端一般放置唐明皇的木偶雕像，因为戏曲界常以唐明皇为行业神，民间称其为"老郎菩萨"。与之并列放置的多为喜神，即舞台上的道具婴儿，在台下也要当作神来供奉，放在大衣箱中起到压箱的作用。大衣箱最上面的第一件戏装是传统京剧中被认为最吉祥的一件，即富贵衣（用于最贫困人物角色的缀有彩色碎绸贴补的戏衣）。虽然破烂，但是在传统剧目中穿"富贵衣"者最终大多金榜题名，化

❶ PAN XIAFENG. The Stagecraft of Peking Opera［M］. Beijing: New World Press, 1995: 111.

身显达富贵。同时，这件最不值钱的衣服放在最上面也保护了下面金绣、彩绣的珍贵戏装不受损坏。依照旧时行规，大衣箱上不准睡觉，箱案不得坐人，不得两脚磕箱；同行来到，需要先到大衣箱前点香吊表，以示恭敬。由此而见，大衣箱在京剧的衣箱制中具有举足轻重的地位，成为京剧舞台演出过程中必不可少的保证。

本节选取大衣箱中颇具代表性的服饰，具体包括蟒、官衣、开氅、鹤氅、帔、八卦衣、法衣、僧衣、褶子、宫装、古装、裙袄裤、旗袍、富贵衣，以及大衣箱中的主要配饰（坎肩、斗篷、蓑衣、领衣、饭单、四喜带、偏带、玉带、丝绦、腰巾、腰箍、云肩、水袖和水裙），并采用英汉文本并列的方式，从不同角度传递文化内涵，关注文化差异。

（一）蟒的文化内涵及其协调翻译

蟒，是京剧衣箱中的必备之物。蟒袍简称蟒，样式为圆领阔袖，右衽大襟，身长及足。腋下有立摆，又称"摆衩子"。蟒为戏曲人物中帝王将相及高官所穿。在清朝，皇室宗亲和一至七品官均穿蟒服。究其本意，蟒即大蛇，比龙少一爪。龙袍上所绣的五爪龙去一爪，即为蟒袍。虽然蟒袍与龙袍略有差别，但依然保持了尊贵的象征。蟒服通身绣龙及吉祥图案，下摆绣海水江崖以及陪衬性的图案。随着身份和级别的差别，袍上所绣龙的数量、颜色以及是四爪还是五爪也有相应的区别。蟒象征着华贵，且由于蟒袍为宽袍大袖，玉带环腰而不束腰，袖端又装有白绸水袖，所以还具备了可舞性。蟒穿在角色身上呈现出静态美与动态美的统一，不同颜色、纹样的蟒服可以表现不同身份和性格的人物。

不同颜色的蟒服内涵不一。红团龙蟒应用的范围很广，为身份高、性格文静的人物所用，如皇叔刘备、驸马陈世美、巡按王金龙等。绿团龙蟒一般为红脸的忠义之士所用，如关羽、关胜、赵匡胤等。以服色之绿与脸谱之红形成"补色对比"，是中国传统的配色方法，此类人物大多智勇双全、文武全才。蟒水纹样一般为直立水，只有关羽穿用的"蟒"选用光泽度极高的全套卧水江崖，更增强了人物的雄奇伟岸之感。黄团龙蟒为中国封建社会里帝王专用，是至尊至贵、皇权的象征，绣活多采用俊雅清丽的绒绣，弯立蟒水。白团龙蟒一般用于英俊儒雅的青年武将（如武小生行当的周瑜）或正直英武的中年武将（如老生行当的杨延昭），给人以潇洒清秀之感，同样多采用俊雅清丽的绒绣，弯立蟒水。凡穿蟒的人物，加用靠领（三尖领）可表示此人是武将。黑团龙蟒的黑色是具有庄重感的中性色。远在周、秦、汉时代，古人因崇拜天而崇尚象征天的黑色，且天子冕服用的就是黑衣。京剧中庄重气派、刚直性格的人物规定服色为黑色。此外又规定，凡勾黑色脸谱、性格粗犷豪放的人物，也以黑作为服色。前一类人物如文臣包拯，后一类人物如猛将张飞、项羽、焦赞、尉迟恭等。

除龙纹外，蟒服前后心的显著位置都会环摆八宝纹样（宝珠、方胜、玉磬、犀角、

古钱、珊瑚、银锭、如意等）。此外，还在全身插底散摆八吉祥（指花、罐、鱼、肠、轮、螺、伞、盖八种饰有风带的吉祥物，又叫佛家八宝）。吐水大龙蟒：蟒上龙纹最生动、最有气势的当属大龙。其姿态一般为龙头朝下、龙尾向上（甩到左肩部位），气势磅礴。依其姿态可称为降龙。龙口喷吐海水，更增生动之感。蟒水中又绣有一条小龙，与之上下呼应。这种大小龙呼应的图案纹样，名曰"教子升天"。全身平金绣，服色为红，象征人物位高权重。团凤女蟒：女蟒的款式与男蟒基本相同，但也有自己的独特之处。首先，它的尺寸短（一般一米左右），长仅至膝部，身后无摆。其次，所绣纹样主要是凤凰、牡丹，以鸟中之王、花中之魁象征至尊至贵的女性。穿用时上身配以云肩，挂玉带，下身系裙。女蟒的色彩一般只用黄、红。黄色是供皇后、贵妃专用，红色用于王妃、郡主（有时也用于公主）。此种女蟒，绒绣团凤，周身单镶黑宽边（波线式），以平金绣云、绒绣仙鹤。行龙女蟒：此种女蟒在纹样上更接近男蟒，绣行龙，下部有蟒水（三江水），如清人所绘《戏剧画册》的《回龙阁》剧中的王宝钏，即用此行龙女蟒，当为早期样式，近世已不多用。老旦蟒属于有很高身份、地位的老年妇女（老旦行当）的专用蟒，其纹样一般用团龙或团龙凤，而不单用凤，有蟒水。在使用上与女蟒的区别在于不配云肩、玉带，而是腰系丝绦，挂朝珠，下身系墨绿大褶裙。这种人物造型比较庄重、沉稳。老旦蟒的色彩一般只用黄和秋香色，黄色用于太后，秋香色用于老郡主、老诰命夫人。

下面以英汉对照版本中关于蟒的介绍为例，全面理解对外传播中蟒的文化蕴意。

原文 The imperial format of embroidered patterns, a terrestrial composition with waves, clouds, mountains and dragons described above, appears in traditional Jingju costumes on the Mang, the most important garment at court for both male and female characters. As the precise duplication of the emperor's robe was prohibited onstage, the decorations were modified. Water is a dominant feature of the Mang design, covering as much as the bottom third of the garment. The embroidery of the waves is particularly impressive as these large areas are solidly filled with stitches. The standing waves can be rendered in rows of satin stitch using three or more gradations of color. The concentric semicircles above have similar gradations in color to define the shapes. An even grander effect comes from couching the entire water area in gold threads, separated by a contrasting color thread. The simplest version has curved standing water made with wavy diagonal lines in a single chevron in the center front and back of the robe, topped with a series of concentric semicircles to indicate still water. More complex patterns employ more chevrons or a greater proportion of semicircles.

On some Mang, dragons play in the waves. The mountains are satin stitched in straight lines, using gradations of colors of thread, or solidly couched. The rest of the surface of the Mang around the dragons may be left plain or have decorative clouds incorporated into the composition. Clouds take on several shapes, either round and fluffy or long and narrow. In addition to the terrestrial design, the dragon comprises a major component of the composition on the Mang, as the dragon has long been an important symbol in Chinese culture.[1]

拙译 京剧中男女角儿最重要的服饰就是蟒袍。蟒袍上有波浪、云彩、山峦和龙的图案组合，为了使其有别于现实版的龙袍，京剧中的蟒袍在细节上有所调整。蟒袍下端斜向排列的线条称为水脚，是蟒袍最显著的特征，其面积占据了整个袍服底部的三分之一，针脚稠密的水脚上有波涛翻滚的水浪，水浪之上又立有山石宝物，俗称为"海水江崖"，即蟒水。海水有立水、平水之分。立水指袍服下摆条状斜纹所组成的潮浪；平水指在江崖下面鳞状的海波。海水意即海潮，潮与朝同音，故成为官服之专用纹饰。江崖，又称江芽、姜芽，如明代官吏织金蟒袍，山头重叠，似姜之芽，除暗示吉祥绵续之外，还寓有国土永固之意。蟒水的具体形状有弯立水、直立水、立卧三江水、立卧五江水、全卧水五种。蟒水的规范性也很强，使用何种蟒水，根据人物的类型而定。蟒袍上立水纹样的挑绣图案采用三种或三种以上的颜色，一般通过渐变色来框定两个同心半圆图案形状。同时，采用对比鲜明的金银线（复合色）刺绣表现整体水面。在蟒袍前后中心位置最简单的刺绣图案是人字形花纹的弯曲立水，上面绣有同心半圆图案来暗示平静的水面；其他复杂的图案则使用更多的人字形或更大比例的同心圆。

在另外一些蟒袍上，还绣有蛟龙戏水。通过使用渐变色的直线或向下弯曲的线条将山的图案绣在缎面上。蟒袍上围绕龙的可能是平原，也可能是与之融为一体的装饰性的云。云的形状各异，有毛茸茸圆形的云，也有长细条的云。除去这些辐射状的图案设计外，龙在蟒服中占据主导地位，正如我们中华传统文化中的龙一样重要。

原文 The ceremonial robe, dubbed Mang or Mangpao in ancient China, was formerly restricted merely to royalty and nobility in real life and was introduced to the stage no more than two and half centuries ago. It was Emperor Qianlong (reigned 1736-1796) who ordered the introduction after he found the robe to be in use in a local Kunqu Opera presentation during a trip south of Yangtze River. I due course, the robe became popular even in the capital, where designers have dramatically enriched it in size and color. Finally, it has been adapted to various dramatic figures, ten colors, half of which are grouped as the five primary colors (Shangwuse), namely, red, green, yellow, white and black, in contrast to the other subgroup of pink, purple, pale-brown,blue and pale-blue, all of which are labeled the secondary colors (Xiawuse). The robe is usually about 1.5 meters long with a large lapel and a round collar as well as a back hem. The front and the back are both embroidered with dragons and flowers, and the hem and the cuffs with waves, all in

---

[1] BONDS ALEXANDRA B. Beijing Opera Costumes: the Visual Communication of Character and Culture [M]. Honolulu: University of Hawaii Press, 2008: 108.

silk thread. Tradition has it that formality seems everything to Peking Opera performers and they would, particularly in former times, choose the right, albeit ragged, robe than don any nice but in male and female, old and young, even as menial as eunuch. The yellow ceremonial robe with a pattern of a coiled dragon (Tuanlongmang) is limited to the part of an emperor. Roles of high-ranking officials, either civil or military, feature black robes with different designs. Bao zheng, a Song minister, has been invested, so to speak, with a black ceremonial robe decorated with a sitting dragon (Zuolongmang) and the third century soldier of fortune Zhang Fei wears a similar robe in undulant dragon embroidery (Xinlongmang). By and large, the role of an old man or Laosheng often calls for a smaller coiled-dragon-pattern robe to show his composure, and that of a "painted face" or hualian usually calls for a larger sitting dragon design on the robe that helps to add to his devil-may-care trait. As a rule, a ceremonial belt, popular known in China as a jade belt (Yudai), must go with a ceremonial robe.❶

拙译 我国古代的礼服被称为蟒或蟒袍，最初仅限于王宫贵族日常生活中的穿着，两个半世纪前才逐步引入舞台。确切地说，应该是在乾隆皇帝在位期间（1736—1796）下江南观看当地昆曲表演时发现了长袍。后来，长袍在京城也逐渐流行，而且其尺寸和色彩在设计师手中也极大地丰富起来。经过改良，适应了不同的戏剧人物，十种颜色中有一半被称为是上五色，分别是红、绿、黄、白、黑，与之形成对比的是下五色，即粉、紫、浅褐色、蓝和淡蓝色。袍服一般长 1.5 米，大圆领镶有黑边，前后都绣有龙和花朵，下摆和袖口还绣有波浪，全部图纹都选用丝线刺绣。京剧表演遵循传统，特别是在过去，所有演员的着装，男女老少，包括太监或仆人，在着装中都坚持"宁穿破，不穿错"的原则。团龙蟒袍仅限于皇帝一人使用，而高级官员，无论是文职还是武将都穿不同图案的黑袍。宋朝的包拯，穿着有坐龙图纹的黑色袍服；而三国时期的张飞则穿着行龙图纹的袍服。总体而言，老人或老生常需要穿盘龙图纹的袍服来体现从容淡定的性格特点；花脸则常需要穿大一些的坐龙图纹的袍服来平添其玩世不恭或无所顾忌的个性。作为规矩，礼服的腰带，即在我国盛行的玉带，大多与正式袍服相配。

1. 蟒的文化内涵及其协调翻译

基于上述关于蟒的总体阐释，聚焦蟒的主要类别，男蟒（Male Cere-monial Robe）、女蟒（Female Ceremonial Robe）以及其他蟒服，将国内外三位学者（赵少华、谭元杰、亚历山德拉）对于蟒的阐释进行类比呈现，并给以简要剖析。为方便描述，采用赵、谭和亚历山德拉分别对其文本给以指代，且此简称贯穿于全书始终，具体如下。

原文 Robes embroidered with dragons are the ceremonial dress of emperors, kings and

❶ PAN XIAFENG. The Stagecraft of Peking Opera［M］. Beijing：New World Press，1995：111–112.

senior civilian. Dragons, phoenixes and flowers are embroidered on satin robes, and their lower hems and cuffs are embroidered with waves crash and under break sun and moon. The robes' colors are divided into the "upper five" colors: red, green, yellow, white, and black, and the "lower five": pink, light green, pale blue, purple and bronze. The colors denoting different ranks. For example, an emperor wear a yellow robe, imperial envoys wear red dragon robes, high-ranking military officers wear green robes, and so on.[1]

拙译 蟒服是帝王将相穿的礼服，华美富丽。其样式为圆领大襟，缎子上绣有龙凤、花朵等图案，下摆及袖口绣有海水江崖和日月。蟒的颜色同样分为上五色（红、绿、黄、白、黑）和下五色（粉、湖、蓝、紫、香），每种颜色由相对应身份的人物穿着，例如，皇帝穿黄蟒、大臣穿红蟒、武将穿绿蟒等。

原文 Mang is a kind of Ceremonial Robes for Emperors and high-ranking officials at the official occasions. It is a cross-over closing robe with a round collar, embroidered elaborately and made of silk. Mang is vivid in color, different colors have their own special symbolic meanings and implication. Mang has both wide robe and wide sleeves, which is easy to perform on the stage.[2]

拙译 蟒服源于明清时期，是帝王将相等高贵人物通用的礼服。圆领大襟，右衽阔袖，袍长及足，并有摆衩子，缎面的蟒袍上绣有海水江崖等图案，极具装饰效果。颜色丰富，分为上五色和下五色，并根据人物形象的设定而具有特定寓意。蟒服有着宽松的腰身和可任意舞动的水袖，并以此传达人物情感。

原文 The name Mang, meaning python, refers to the four-clawed creature embroidered on the court robes in the Ming and Qing dynasty. Emperors and high-ranking members of the court wear the Mang for official scenes. Mang had a rounded neckline and back pieces were continuous, without a shoulder seam. Full length, tubular sleeves with water sleeves. As the most prestigious costume, the Mang has rich, symbolic designs embroidery in solid gold or beautiful colors. The embroidery such as dragons in a terrestrial environment of waves, mountains and clouds. The use of color in Mang exemplifies the system of upper (Shangwuse) and lower colors (Xiawuse). The more pure upper colors are used on the higher-ranking characters, and the lower colors are less emphatic on stage. Generally the Mang is worn with a hop shaped jade belt that encircles the body, and with padding (Pang'ao), a white collar (Huling), red trousers and high-soled, high-topped boots.[3]

❶ 赵少华. 中国京剧服饰 [M]. 北京：五洲传播出版社，2004：8.
❷ 谭元杰. 中国京剧服装图谱 [M]. 北京：北京工艺美术出版社，2008：17.
❸ BONDS ALEXANDRA B. Beijing Opera Costumes: the Visual Communication of Character and Culture [M]. Honolulu: University of Hawaii Press，2008：115.

京剧服饰大衣箱文化阐释及翻译协调美 第三章

59

拙译 蟒的原意是蟒蛇，属于明清时期宫廷长袍上绣制的四爪动物。蟒袍是皇帝和高级官员在朝廷上的着装，圆领口，后片相连，没有肩线，全身长，水袖呈管状。蟒袍作为最高规格的服装，其丰富而极富象征性的设计图案全部采用纯金线或色彩美丽的金线绣制，如处于波浪、山峦和云朵中的龙纹。蟒袍在颜色的使用上遵循上五色和下五色的用色规矩。高级别官员的蟒袍更多使用上五色中的纯色系，舞台上的下五色则用于相对一般的官员的袍服。通常袍服上都要配一个环状的玉带，环绕腰部，并配有垫物（胖袄）、白领（护领）、红裤子以及厚底高帮靴。

译文剖析 京剧服饰蟒的字面意思是大蛇，比龙少一爪。龙袍是帝王所穿，而蟒则为比帝王级别低的人穿用。亚历山德拉通过"蟒蛇，四个爪子的动物"（python, the four-clawed creature）凸显了这一文化内涵，而赵少华与谭元杰的版本则对此没有介绍。

对于蟒的定位与穿着场合的阐释，赵少华译为"the ceremonial dress of emperors, kings and senior civilian"，《牛津高阶英汉双解词典》对"civilian"的解释为"a person who is not a member of the armed forces or the police"❶，指代与军人、警察相对的平民百姓，这在一定程度上有悖于身份高贵的官员或皇室宗亲才能穿蟒的定位，可见选用"senior civilian"阐释并不准确，且忽略了蟒只有在正式场合穿用的规定。谭元杰对于蟒的阐释为"emperors and high-ranking officials at the official occasions""high-ranking officials"，虽表明蟒为高级别官员所用，但并未强调只有皇帝以及皇亲国戚才有资格穿用。而在亚历山德拉的表述"Emperors and high-ranking members of the court wear the Mang for official sense"中，仅用"the high-ranking members of the court"就简洁明了地阐释了穿用人物的身份。根据《牛津高阶英汉双解词典》对于"the court"的解释，"the king or queen, their family, and the people who work for them and give advice to them"❷（皇帝、皇后等宫廷或王室成员），较好地限定了蟒袍的穿用场合与人物身份。

对于蟒的形式、面料、颜色与纹样，三位学者阐释各异。赵译忽略了蟒的形式，虽借助"a cross-over closing robe with a round collar"说明交叉叠领，但并未指明右衽与左衽的细节；谭译对此也只是一带而过；亚历山德拉阐释的"rounded neckline, back pieces were continous, without a shoulder seam, full lenghth, tublar sleeves with water sleeves"指明了圆领特征，但忽略了大襟特色。笔者认为关于领子的描述可以协调翻译为"the robe has a round collar and broad lapels"，避免读者对"右衽"这一中国传统服装术语产生歧义。

对于蟒的长度描述，亚历山德拉用"full length"（长及全身至地面），但实际其长及足，因此阐释为"anklelength"（袍长及足）更为准确。赵、谭两位则对此只字未提。

同时，亚历山德拉对于蟒服袖子的描述为"tublar"，未能表达其"阔袖"的特色。

❶ 霍恩比. 牛津高阶英汉双解词典［M］. 王玉章，赵翠莲，邹晓玲，等译. 7版. 北京：商务印书馆，2010：345.
❷ 同❶459.

谭译虽用了"wide robe with wide sleeves"（宽松的宽袖袍服），但并未表明其袖口较袖窿更宽。如若表述为"wide cuffed with water sleeves"（带水袖的阔袖），可以更好地阐释其特点。

针对蟒的面料，赵选用"satin"（绸缎）一词，将其与服饰图案，从中心的龙、凤与花朵到底部的海水江崖，一并描述；谭译选用了绫罗绸缎的总称"silk"（丝绸），较好地再现了蟒的材质。至于针对图纹的描述，赵译采用了解释性的表达"waves crash and under break sun and moon"，帮助读者理解其图纹的构成；亚历山德拉则通过"dragons in a terrestrial environment of waves, mountains and clouds"凸显龙、海浪、山峦和云朵。可惜三位学者似乎都忽略了"八吉祥、暗八仙、云头等吉祥图案陪衬"的介绍，如若表达为"there are some lucky designs such as The Lucky Eight embroidered on the robe"，则在一定程度上更方便西方读者的理解。

针对蟒的颜色表述，赵和亚历山德拉都选用了"upper five color and lower five color"，但是并未阐释其具体内涵，特别是颜色对于身份地位的象征等；谭用"vivid in color"概括蟒的颜色，也同样没有忠实地传递蟒的服饰颜色内涵。

总体而言，对于蟒的英文表达，三位学者各有千秋。赵与谭省略了一些容易使西方读者产生歧义的细节，侧重于蟒的总体介绍，亚历山德拉则似乎更关注蟒所蕴含的传统文化内涵。

2. 女蟒的文化内涵及其协调翻译

女蟒的面料、颜色与男蟒一致，花纹多为凤与牡丹，并配上其他的吉祥图案，一般为后妃、公主、贵妇或女将穿着。因角色需要，颈部可围云肩，腰系玉带或丝绦，内穿裙子。款式为齐肩圆领，阔袖并带水袖，长及膝，腋下无摆，大襟扎带。

原文 The female ceremonial robes need to be dressed with skirts. Normally the patterns are phoenix and peony.[1]

拙译 女蟒需要和裙子搭配穿用，通常绣有凤和牡丹图案。

原文 Queens and woman generals also wear dragon robes. They are much the same in design as male dragon robes, but shorter , reaching the knees and without the rear hems.[2]

拙译 女蟒为后妃等女性角色穿用。其款式设计与男蟒相同，但长度及膝且无后摆。

原文 The Nümang has two cuts, one with straight edges and a traditional shape, and another with an additional border.The straight-edged Nümang reflects the men's form, only in

---

[1] 赵少华. 中国京剧服饰 [M]. 北京：五洲传播出版社，2004：8.
[2] 谭元杰. 中国京剧服装图谱 [M]. 北京：北京工艺美术出版社，2008：18.

a shorter length. This kind of Nümang is worn by mature women. And the newer design has an attached border of contrasting fabric with wavy edges around the hems and side seams. The wavy-edged Nümang appears on young women. The colors are fewer options for the fabric of women's Nümang than men's Mang. The dragon ornament of Nümang have only eight, less than men's Mang of ten dragons. And there are flowers on the surface. The straight- edged robe is designed with dragons, but the wavy-edged Nümang has phoenix designs, which represents the empress.❶

拙译 女蟒分为两种，一种是传统的直边款式，另一种是多边款式。直边的女蟒类似男款，只是在长度上要短些，这种女蟒属于成熟女性的专款。较新的设计款式是在下摆以及侧缝处都采用对比色调鲜明的微波浪状的宽边面料镶嵌。但是，这种波浪状褶皱边的款式一般为年轻女性选用。女蟒与男蟒相比，其面料的选择余地较小，即使是有龙的装饰图案的女蟒，龙的数量也只有8条，与男蟒的10条龙的图案相比也要少2条。此外，男女蟒袍面料上都有花卉图案。直边的龙袍边缘设计有龙的图案，而女蟒作为皇后的象征，边缘则有凤的图案。

译文剖析 女蟒为宫廷中女性角色穿用，赵译阐释了女蟒与蟒的关系（"there are two styles of the ceremonial robe. One is for male, the other is for female"）。而谭译"Queens and women generals wear dragon robes"则阐释了女蟒的龙纹图案，但关于穿用的人物身份为皇后和女性角色等表达则略欠妥帖，可以协调为"queens and high-ranking female members of the court"，使女性角色具体表述为"后妃、公主、贵妇或女将"等尊贵身份，避免误解为普通女性。由于女蟒和蟒的式样大致相同，对女蟒形制的阐释，谭译和赵译都将其简化了，只关注女蟒长度至膝处，没有后摆。对于女蟒穿用时，需穿内裙、颈系云肩、腰佩玉带系列等具体要求，赵译只是说明了需搭配裙子（"needed to be dressed with skirts"），并未传递该裙穿在女蟒内等信息。而谭译忽略了穿着女蟒的特别方式，但对于男、女蟒的形式借助"same in design, but shorter, reaching the knees and without the rear hems"给予了相关表达。亚历山德拉介绍了女蟒的长度，但并未使之具体化。综合三位学者的表述，不妨将其协调翻译为"same in design, but knee-length, carries no hemline at the back, need dressed with a cloud shawl（Yunjian）around the shoulders, a jade belt around the waist and a skirt inside the robe"。

此外，亚历山德拉在对女蟒款式进行介绍时，关注到了与蟒大致相同的传统款与带边的新式款（"there are two kinds of Nümang: one with straight edges and a traditional shape, and another with an additional border"），阐释了新式样的女蟒在下摆和边缘处因使用不同颜色的面料与波浪式的褶皱，导致其与服饰本身反差明显；甚至还阐释了相比较传统的女蟒，新型女蟒更适用于年轻女性，可谓观察的细致入微。

在表达女蟒的颜色与面料时，由于其颜色和面料同蟒一样，所以赵和谭并没有对

---

❶ BONDS ALEXANDRA B. Beijing Opera Costumes: the Visual Communication of Character and Culture [M]. Honolulu: University of Hawaii Press，2008：118.

此给以阐释。但是亚历山德拉关注到女蟒的颜色在选择上少于男蟒。而在针对纹样方面，赵和亚历山德拉两位都提到"phoenix and poeny"，但是前者描述更为详尽，包括龙的数量和位置，以及凤的象征意义等，而谭对此全部忽略。

3. 旗蟒的文化内涵及其协调翻译

满族人因隶旗籍，所以又被称为旗人。旗蟒原为清朝满族皇后所穿的吉服——朝袍，后经艺术化处理成为京剧服饰，通用于中国古代少数民族统治阶层的贵族妇女。旗蟒款式为捻襟（大襟）圆领，长袖，袖端缀马蹄袖口，并绣有团龙和云纹等图案。使用时，内衬领衣儿，外挂朝珠。头饰为大拉翅，足穿高底鞋（鞋底状如倒置的花盆，高6~10厘米，俗称花盆底鞋）。

原文 Qimang is a kind of ceremonial robe for the queen. In Peking Opera, though Qimang was a kind of Manchu court robes , now are worn by non-Han women onstage.[1]

拙译 旗蟒为清朝皇后所穿的吉服。尽管在京剧中旗蟒属于满族宫廷服饰，但现在为京剧舞台中所有非汉族女性穿用。

原文 Qimang is a kind of ceremonial robe for minority noblewoman.[2]

拙译 旗蟒是少数民族贵族女性穿用的礼服。

原文 Qimang resembles the dragon robe worn by Manchu noble women late in the Qing dynasty. It is worn by high-ranking ethnic groups other than Han Chinese. It has a round neck, an S—curved closing that crosses over to the right, tubular sleeves but does not have water sleeves. The ornamentation of Qimang are the same as man's Mang, only eight dragons. The color of such ethnic robes are different from Han people, they often use blue or red colors. When wear Qimang, often wear with a separate blue collar (LingYi) under the gown.[3]

拙译 旗蟒与清代晚期满族女性穿的龙袍相似，只有少数满族高官能够穿用，汉族不穿。领口为S型的右衽交领，袖型为管状但没有水袖。旗蟒的装饰图案与男蟒相同，只有8条龙。这种少数民族的袍服不同于汉民族，经常使用蓝色或红色。穿旗蟒时，在袍服下面还配有独立的蓝领（领衣儿）。

译文剖析 旗蟒作为蟒袍的一种，原为皇后穿着，成为戏服后为少数民族贵族女性穿用，针对其身份标识，三位学者给出了不同的表达（赵："was a kind of Manchu

---

[1] 赵少华. 中国京剧服饰［M］. 北京：五洲传播出版社，2004：22.

[2] 谭元杰. 中国京剧服装图谱［M］. 北京：北京工艺美术出版社，2008：50.

[3] BONDS ALEXANDRA B. Beijing Opera Costumes: the Visual Communication of Character and Culture ［M］. Honolulu：University of Hawaii Press，2008：122.

court robes, now are worn by non-Han women onstage";谭："is a kind of ceremonial robe for minority noblewoman";亚历山德拉："it is worn by high-ranking ethnic groups other than Han Chinese")。笔者以为，原文"non-Han"的表达对于非汉语为母语的读者而言，可能会对"Han"的理解产生歧义；查阅《牛津高阶英汉双解词典》，"minority"意为"a small group within a community or country that is different because of race, religion, language, etc."[1]（国家或团体内因为种族、信仰或语言等差异而区别于与别人的小部分人），强调了数量的少以及其他各方面的差异，因此"minority noblewoman"对于少数民族贵族女性的表达在一定程度上有失尊重，容易使外国读者产生误解；亚历山德拉表述的"ethnic groups other than Han Chinese"指排除了我国汉民族以外的其他民族，但同样也忽略了"ethnic"潜在的歧视。在当今语境下，对于我国少数民族的合理表达，学界的观点是"ethnic minorities"。

同时，赵和谭并没有将旗蟒的服饰形式、面料和色彩等文化内涵充分表达出来。亚历山德拉强调了圆领右衽、长袖无水袖的特征，但对于其特色的马蹄袖并没有关注。笔者认为改用"tublar sleeves with horseshoe-shaped cuffed"可忠实地传递其服饰内涵。旗蟒的图案有团龙和云纹等，亚历山德拉只阐述了龙纹文化。对于云纹等其他图案，不妨借用"as same as man's Mang"，暗含龙纹图纹外还有云纹、海水江崖等图案。针对实际穿用中内衬领衣儿、外挂朝珠的具体规定，亚历山德拉表达了"when wear Qimang, often wear with a separate blue collar under the gown"，但忽略了有关朝珠的表达。笔者认为可以协调补充翻译为"under the gown and with court beads"，暗指宫廷或王室成员佩戴的珠子，相对全面。

4. 改良蟒的文化内涵及其协调翻译

"改良"二字属于清末民初词汇，具有褒贬双重含义。凡对传统事物有所变革，就称为"改良"，时至今日，"改良"重在强调传统基础上的创新。京剧服饰改革，早在19世纪初期就在我国南方的京剧界率先推行，北方杰出的表演艺术家马连良先生遥相呼应，改良蟒由此诞生。其式样和尺寸与男蟒一样，只是在图纹上保留传统男蟒主要部位——前胸、后背的蟒纹，以及袖口、下摆的海水江崖，删去了流云、八吉祥等插底纹样，重新设计了蟒纹图案，使之变得清爽别致。同时，在团龙的下端，不拘一格地采用了行龙纹样，形成了团龙和行龙融为一体的图纹。团行龙改良蟒一般用于老生行当的角色，服色一般为淡雅沉着的秋香色或浅驼色，以象征人物之老练凝重。关于改良蟒发展历程与使用效果的英文表述和汉语译文概述如下。

原文 By the end of the 1930s, a simplified model of man's robe came into being as a result of improvements made by both Zhou Xinfang and Ma Lianliang, two superstars of male roles

----

❶ 霍恩比. 牛津高阶英汉双解词典［M］. 王玉章，赵翠莲，邹晓玲，等译. 7版. 北京：商务印书馆，2010：1280.

in Peking Opera. This fashion generally known as the improved robe (Gailiangmang) is devised more plainly than the traditional one, with a breast dragon in embroidery and some embellished waves on both the hem and the cuffs. The change has mad it easier both to put on and take off the garment. Ma is said to have slipped into such a robe when he was acting Qiao Xuan, an elder statesman of Wu, in *Ganlu Temple*.❶

拙译 20世纪30年代末，改良版的男袍服经过周信芳和马连良两位京剧名角的改进，终于问世。这种改良蟒袍比传统袍服更加时尚，胸部绣有龙的图案，下摆和袖口绣有波浪，并且方便穿脱。正如马先生所说，在扮演《甘露寺》中的乔玄、吴国太时，很容易就套上了这改良版的行头。

无独有偶，赵少华与美国的学者亚历山德拉对待改良蟒也有着类似的阐释。

原文 Gailiangmang is a kind of dragon robe modified by Ma Lianliang. He noted the double-dragon design on the chest, simpler than conventional Dragon Robes, but more attractive.❷

拙译 改良蟒是由京剧大师马连良改进的龙袍，将团龙蟒的十个团龙简化为前后胸各绣一个大面积的团龙，并删去了流云等搭配图案，突出了胸部的双龙，比传统龙袍更为简洁且更具魅力。

原文 Description and History. The introduction of the Gailiangmang is attributed to Ma Lianliang (1901−1966), who was interested in a simpler look from ancient times. The cut of the Gailiangmang is narrower than the standard Mang, and the surface ornamentation has been reduced as well.

Usage.The Gailiangmang can be worn in informal scenes by Sheng in the role of officials, current and retired.

Color and Ornamentation. The Gailiangmang comes in maroon, black, olive green, beige, and other colors in the subdued range. The nature of the surface decoration indicates the level of Importance. High officials may have a single large dragon embroidered on the chest and back or have an additional pair of walking dragons placed above the waves on the front and dragons placed above the waves on the front and back. The simpler smooth dragons from archaic Chinese designs can be used for lower or retired officials. Clouds and other emblems are not depicted; instead, there are bands of embroidery around the neckline and sleeve hems. The waves may be standard size or reduced.

Accessories. Either a jade belt or a fitted cloth belt of matching fabric can be worn with a

❶ PAN XIAFENG. The Stagecraft of Peking Opera［M］. Beijing：New World Press，1995：113.
❷ 赵少华. 中国京剧服饰［M］. 北京：五洲传播出版社，2004：16.

Gailiangmang. High-topped, tall-soled boots are worn as well.

Headdress: The Gailiangmang may be worn with a Zhongsha, or one of the gold filigree hats appropriate for court, such as the Xiangdiao (prime minister's hat).❶

拙译　文字描述和历史渊源：改良蟒在舞台上的应用要归功于马连良（1901—1966），他对自古以来的简化服装颇感兴趣。改良蟒在裁剪上比普通蟒服要贴身，并且其表面很多装饰物也被删减。

穿用场合：改良蟒在京剧中属于生角儿处于在职或离任时，非正式场景中穿用的袍服。

颜色和装饰：改良蟒色彩柔和，采用红褐色、黑色、橄榄绿、米黄色以及其他较柔和的颜色。表面装饰的属性暗示了角色人物身份的重要程度。高官可能会在胸前和背部绣一条大龙，或者在前面的波浪图案中再另外绣制一对行龙，或在蟒服正面和背面的波浪上绣制龙纹。蟒服上简洁流畅的龙纹可用于官位低或离任的官员，而且没有云和其他元素，只是在领口和袖口有刺绣花边，波浪可以是标准尺寸或小号的。

配饰：无论是玉带还是相应的布带都可以和改良蟒配套穿用，搭配厚底高帮靴。

头饰：改良蟒可以和中纱帽匹配，也可以和朝廷中使用的金银丝帽子匹配，如相貂（选用有色缎子，且帽翅两端缀有光珠和龙头的帽子）。

译文剖析　改良蟒来源于蟒，由京剧名角马连良在实践中简化得来。赵的表述突出了改良龙袍的创始人，改良的重点是双龙在胸，这样会比传统龙袍更好看。亚历山德拉则主要描述改良蟒的历史，关注马连良的贡献，强调了其对于龙袍简洁性的关注，相形之下，赵的表述对于"dragon robe"（龙袍）与蟒袍的差异并未说明。同时，亚历山德拉对于"改良蟒"的用途、颜色和图案，包括双龙图案的设计及其具体位置（前胸和后背）都有说明，翔实周到。

在改良蟒中有一款袍服需要特别引起关注，那就是太监所穿的蟒袍（Taijian Robe 或 Eunuch's Robe），又称"老公衣"，属于改良或创新蟒。其特点是圆领大襟，阔袖有水袖，周身彩绣宝蓝色团龙。腰间、下摆、袖口和领口处均加异色阔边，腰系排穗。服色分红、黄两种。穿黄色太监蟒袍的侍奉皇帝；穿红色太监蟒袍的为剧中称王角色的随从。其介绍用英文简述如下。

原文　It is similar in design to the Xuezi. It is made of yellow or red silk with a wide collar, wide lapel and wide cuff with "water sleeves" (Shuixiu). It is decorated with sapphire blue dragon designs and a wide fringe of blue satin is added around the wrists and the collar, whilst the waistline is usually braided.❷

❶ BONDS ALEXANDRA B. Beijing Opera Costumes: the Visual Communication of Character and Culture［M］. Honolulu：University of Hawaii Press，2008：124.

❷ 阙艳华，董新颖，王娜，等. 中英文对照京剧服饰术语［M］. 覃爱东，陶西雷，校译. 北京：学苑出版社，2017：30.

同样，潘霞凤女士也对太监蟒袍给予了关注，其英文表述与笔者的汉译呈现如下。

原文 A very rare type of the ceremonial robe is the so-called eunuch's robe (Taijianmang) that is not included in the ensemble. It was merely a personal design by the famous "painted face" Hou Xirui for his part as Liu Jin, the arch-eunuch of Ming dynasty, in *Temple of Karmic Law*. This crimson robe with a round collar is different from nearly all of its kind. It has both front and back sitting dragon patterns and is edged all over with golden bands nearly seven inches in breadth, while a string of tassels in apricot yellow dangle about waist.[1]

拙译 在系列袍服中，有一种比较少见，那就是太监的袍服（太监蟒）。太监蟒是由著名的花脸演员侯喜瑞在扮演京剧《法门寺》中的刘瑾（明代的大太监）时，针对原有的京剧服饰给以个性化处理和改良的结果。这种深红色的圆领袍服几乎不同于以往任何一种其他款式。袍服的前、后身都有坐龙的图纹，且所有边缘处都镶有将近23厘米宽的金色边，杏黄色的流苏从腰部垂下。

补译 太监蟒袍的基本样式与褶子相同，用黄缎制成，绣有宝蓝团龙，四周有花边，腰际有排穗，多为小太监所穿。而大太监可以直接穿蟒、花褶子等。

拙译 Taijianmang shares similar basic style with Xuezi. Taijianmang is made of yellow satin, and embroidered with blue dragon patterns. It is ornamented with bands around the edges and a string of tassels dangling from the waist. Taijianmang is in most cases the exclusive costume for lower ranked eunuchs. Senior eunuchs of higher social status are authorized to directly wear eunuch's Mang, Huaxuezi and so on.

（二）"官衣"的文化内涵及其协调翻译

官衣是京剧中官员（此外还有个别情况如新科状元、婚典新郎等）穿用的礼服，源于明代官服——盘领窄袖大袍。形制基本与蟒相同，圆领右衽，长袍大袖，袖端加水袖，身后两侧自袖裉以下有摆；材质为素色缎料，不绣纹样，只在胸前与后背各缀一块方形绣品，称为补子，上绣飞禽及旭日海水等图案，有紫、红、蓝、黑、香色等颜色，在明、清时代是区分官阶与身份职务的一种标志。官衣颜色是官阶大小的标志，紫色、红色象征的身份和品位最高，蓝色次之，黑色最低。黑色而无补子的称为青素，为驿丞、门官所穿。一般文官的官衣绣有禽鸟的图案，武官的则绣有走兽，所绣纹样都有严格规定。女官衣较男官衣稍短，袖裉下无摆，除贵妇外，有时丑扮的官员也穿。男官衣穿时腰围玉带，女官衣则束软带或系丝绦。依据款式和使用功能，京剧中的官衣可分为改良官衣、学士官衣和判官衣。

❶ PAN XIAFENG. The Stagecraft of Peking Opera [M]. Beijing: New World Press, 1995: 113.

原文　The official robe (Guanyi) is made for civil servants only and there are both male and female versions. The male official robe has, as a rule, a round collar above a large lapel against a back hem insofar as there are both front and back ornaments in patterns either of a sun, a dragon, or surging waves. Sometimes a crane may take the place of the foregoing. However, the color is crucial as a symbol of official rank in a bureaucratic system, for example, an actor playing the part of a minister is accorded a purple robe, such as Lu Su (172−217) in *Meeting of Heroes*. Next come the red robe for the role of a prefect and the blue one for a magistrate. A black robe with no design is worn only by a janitor, like Yan Xia in *Beating the Wicked Official Yan Song*.

Usually the female's official robe would be much shorter than that worn by a man. Simply tinted in only red and purple, the former is ordinary adopted as a wedding garment and the latter is much used by actress playing old ladies of rank, *Madam Wang Visits Her Daughter' Cave Dwelling*. In addition, a clown playing an oafish magistrate sometimes also wears a red female's robe, as is evident in one of the scenes of *Reunion of Liu Lujing and His Wife*.

In the past, female official robes were customarily chosen for occasions of curtain-call acknowledgments. On the first day of the first lunar month, it is recounted, a new opera for good luck was conventionally put on the stage. In order to extend their gratitude to the loyal theater-goers, two of the players would always appear at the end of the performance and bow. One of them would be an actor playing an old man's role but wearing no beard, and the other, a performer acting the part of an old woman; both were in red female official robes as tradition dictated. A similar ceremony would be necessary when a good piece was rewarded with a gift of money by the host during performances sponsored as well as staged in his residence.

It is noteworthy that close ties could not be ignored between official robes and the bureaucratic system of nine grades through the ages in feudal China. The designs of filigreed ornaments on the dress, what is more, are by far the most substantial because they always serve, apart from decoration, as a sort of coat of arms symbolizing the wear's official rank. It varied on the basis of what form of bureaucracy a dynamic ruler was adhering to. During the Qing dynasty (1644−1911), it was decreed that civil officials should be distinguished by various embroidered birds on the coat, and military officers, by animals. Starting with the design of a crane that stood for the highest civil official rank, the bird patterns came out successively as the golden pheasant, the peacock, the wild goose, the silver pheasant, the egret, the purple mandarin duck, the quail, and the flycatcher. In the same way designs of beats were arranged on military officers with the unicorn in the lead, followed by the lion, the leopard, the tiger, the bear, the young tiger, the rhinoceros, and the sea horse.❶

---

❶ PAN XIAFENG. The Stagecraft of Peking Opera ［M］. Beijing: New World Press, 1995: 114–115.

拙译 官袍（官衣）仅供公职人员使用，男女款式不同。按照惯例，男性袍服为圆领右衽，其前胸与后背各缀一块方形补子，上面绣有飞禽、旭日以及海水的图案，有时也可能会用鹤的图案取代。颜色在袍服中非常重要，成为官阶高低的象征。例如，扮演大臣的演员，《群英会》中的鲁肃（172—217），就身着紫袍。比紫袍低一个级别的是红袍，蓝色是地方官员的袍服。没有设计图案的黑色袍服是守卫的着装，像《打严嵩》中的严侠就只穿黑袍。

通常女性的官袍比男性的要短很多（穿用时下边也穿裙子）。概括而言，只有红色与紫色。红色一般用于婚庆，紫色常被有身份的老年诰命夫人穿用，如《母女会》（又名《探寒窑》）。此外，京剧中扮演笨手笨脚的地方官的丑角有时也穿红色女袍，就像《小上坟》（又名《飞飞飞》《丑荣归》）中的场面一样。

过去，女性官袍通常在"金榜谢幕"或"金榜谢赏"时穿用。在农历正月初一通常要上演新的吉祥戏以图吉利。在表演结束时，剧院会选派两名演员（一名不戴髯口的老生和一名老旦），全都身着传统的红色女官衣登台向忠实的观众表达感谢。当一部好的京剧作品获奖时，投资人在其首演日举办这样的仪式也同样非常必要。

值得注意的是，在中国封建社会，九级官衔与袍服之间的密切关系不容忽视。而且袍服上的金丝装饰非常关键，除了装饰功能之外，更是官衔等级的标志性象征，这一切都基于不断更换的统治者对于形制的坚守。清朝（1644—1911）是以补子纹样区分官阶与身份的，所绣纹样皆有严格规定，文绣飞禽，武绣走兽。按照清朝的制度，文官官衣的补子上，一品绣仙鹤，二品绣锦鸡，三品绣孔雀，四品绣云雁，五品绣白鹇，六品绣鹭鸶，七品绣鸂鶒，八品绣鹌鹑，九品绣练雀。武将则是用野兽的图案，一品绣麒麟，二品绣狮子，三品绣豹，四品绣虎，五品绣熊罴，六品绣彪，七品、八品绣犀牛，九品绣海马。而且哪一等官级用哪种补子都有严格的规定，不能乱用。

原文 Guanyi are used for civil and military officials below prime ministers. These are use white satin, and not embroidered with dragon and cloud patterns. However, on the chest and back of such a robe there is "补子" (Buzi) — a rectangle embroidered with a pattern denoting rank.❶

拙译 官衣是文武官员所穿的礼服。素缎上不绣云、龙，只在前胸和后背各绣一块方形的补子。补子是明清时代区别官员级别职务的一种标志。

原文 The Guanyi comes from the official robe worn by officers of the court in the Ming dynasty. When the emperor and the uppermost men of rank at court wear Mang, the middle-ranked civil and military officers appear in Guanyi, though military Guanyi are rarely seen. It has a round neckline and curved crossover closing to the right. Loose sleeves with water sleeves attached.

---

❶ 赵少华. 中国京剧服饰［M］. 北京：五洲传播出版社，2004：28.

The fronts are finished at the side seams with extensions made from a contrasting color. The Buzi (rank edge), used in the Ming and Qing dynasties, appears at the front and back of guanyi at chest height.❶

拙译 官衣来自明朝宫廷官员穿的袍服。当皇帝和宫中最高级官员身着蟒服时，中等级别的文武官员则选用官衣，不过武官身着官衣的情形很少见。官衣的款式一般为圆领、右衽，并带有水袖。其前片侧面的延伸部分还会采用对比色调。明清时期的补子（官员等级的标志）通常出现在官衣齐胸的前后处。

译文剖析 在上述两个针对官衣的英文介绍中，后者显然更为翔实。首先，赵译关于"Guanyi are used for civil and military officials below prime ministers"的表达需要在最后补加"in ancient China"的限定，否则没有时间轴的约束，容易带来概念上的混淆。其次，官衣是源自明朝的官服，而"prime ministers"的表述方式与亚历山德拉阐释的"…the uppermost men of rank at court wear Mang, the middle-ranked civil and military officers appear in Guanyi"相比，后者更全面且易于英语读者理解和接受。

同样，对于官衣颜色的表述，赵的表述"white satin"（白色绸缎）略欠妥当，因为官衣实际应该是素色绸缎，并非白色。如果改为"plain colored satin"（素色绸缎）则更贴近原文本意。对于官衣中补子的阐释，赵译的解释性表述如果在"pattern"后补加"such as birds and animals"则更为详细。总体而言，赵与亚历山德拉两位学者对于官衣的表述，后者更为直观、综合与立体。下面将在一定程度上更为详尽地描述和介绍官衣。

1. 改良官衣的文化内涵及其协调翻译

对官衣加以美化，采用平金绣的圆形纹样取代补子，并在袖口及腰部以下适当饰以纹样。此种改变了图案的官衣被称为改良官衣，为近世所创。

原文 The Gailiang Guanyi has the same cut as the conventional Guanyi, but instead of having a square of embroidered fabric stitched on to indicate the rank, the Gailiang Guanyi has a round motif embroidered onto the robe. One or two additional bands are embroidered parallel to the hem at the height of the knee, and the neckline is encircled with embroidery.

The Gailiang Guanyi is worn at court by Laosheng or Chou, depending on the nature of the character.

The Gailiang Guanyi comes in the colors of the standard Guanyi: red, blue and black. The embroidery usually includes an early form of the dragon with a smooth body in a round design.

A hooped-shaped jade belt and high-topped boots are worn with this garment, along with the

❶ BONDS ALEXANDRA B. Beijing Opera Costumes: the Visual Communication of Character and Culture [M]. Honolulu: University of Hawaii Press, 2008: 147.

inner collar, trousers, and the padded vest used only for the Sheng roles. The Gailiang Guanyi is worn with the Zhongsha, the same headdress worn with standard Guanyi.❶

拙译 改良官衣与传统官衣的剪裁方法一致，但是没有方形刺绣补子来标识官衔级别，取而代之的是袍服上的圆形图案。一两条附加的刺绣缎带长及膝处，领口四周都有刺绣图案。

改良官衣在朝堂上由老生或丑角穿用，这完全取决于人物的性格。

改良官衣的用色源于标准官衣的颜色，红、蓝或黑。刺绣图案通常包括早期龙的形态，其设计通体光洁。

改良官衣的配饰包括官袍上环形的玉带和高方靴，再加上仅限于老生穿用的内领、裤子和有垫肩的马甲。此外，改良官衣通常和中展纱帽一起穿用，头饰标准与官衣无异。

2. 学士官衣的文化内涵及其协调翻译

学士官衣属于官员、学士的常礼服。它是在缀有补子的青官衣基础上演变而成的。身后仍有摆，但不挂玉带，改系软带，软带正中垂下两条飘带。以纹样进行局部装饰，绣边，下摆绣海水。学士官衣是从其母体——官衣中分离出来的，而且依旧未脱离官衣的基本形制。其详细介绍可参阅下列对照文本。

原文 The Xueshiyi is worn by well-read and well-educated young men of no special status or rank. The rounded neckline and curved crossover closing of the stage Xueshiyi reflect Ming traditions, while the hem that curves up on the side vents has a theatrical basis. The Xueshiyi is usually worn with a tight belt of matching fabric, with two tabs hanging down at the center front. The sleeves are tubular with water sleeves. The robe is softer, like the Xuezi, rather than resembling the stiffer robes of court officers. The Xueshiyi comes in black, blue and white. Black reflects the lowest rank of the scholar officials. The ornamentation on the Xueshiyi includes borders around the edges and a medallion at the chest. The border at the intersection between the hem and side vent curves to follow the contour of the outer edge.❷

拙译 学士衣是那些没有特殊头衔或官职的爱读书且受过良好教育的年轻男子之专属。在舞台戏曲中，交叉圆领的款式体现了明朝的传统。侧面开衩，摆部镶边，这些都遵循戏曲服饰的规矩。学士衣通常采用材质相配的带子系紧，有两条带子从袍服的前片正中垂悬而下，并带有水袖。袍服柔软，类似褶子，有别于硬挺的官服。学士衣一般为黑色、蓝色或白色，其中黑色用于最低级别的学士。摆衩的边缘和侧面的交接处需顺着外边沿的轮廓呈现出弧度。

❶ BONDS ALEXANDRA B. Beijing Opera Costumes: the Visual Communication of Character and Culture ［M］. Honolulu: University of Hawaii Press，2008：148–149.

❷ 同❶150.

参阅国内资料，整理后发现还有类似两段有关学士衣的英文表述：

原文 Xueshijin (scholar's kerchief), also known as Jieyuanjin (Jieyuan 即 "解元", is a scholar China who won the first in provincial imperial examinations). It is scholars in the Xueshiyi may wear a Xueshijin. (scholar's hat), a soft hat is made of satin and embroidered with flowers, with the color and pattern used to determine by the rest of the costume. The cap has a tilt, so that the front part sits lower than the back part. It has two Ruyi-shaped (Ruyi, an S-shaped ornamental objects symbolizing good lock) soft wings attached to the back. ❶

拙译 学士巾又名解元巾，解元是中国古代在国考中荣获全省第一名的考生。穿学士衣的学士一般都戴学士巾（学士帽），这种软帽的材质为缎料，样式为前低后高，帽顶左右两端倾斜翘起，帽背上插一对如意形软翅，分花、素两种，使用时与人物服装颜色相配合。❶

原文 The scholar's garment is much the same as a black official robe, only with a design tacked on and both the hem and edges in embroidery, as is worn by Zhu Suiliang❷ in *Raising Ten Memorials in Succession*. Ma lianliang used to wear a red scholar's garment after Kou Zhun's promotion in *Kou Zhun interrogates Pan at Night*.❸

拙译 学士衣与黑色官袍类似，只是在开衩与边缘处都有缝制的刺绣装饰。在京剧《十道本》中，褚遂良身着的就是学士服。演员马连良在《审潘洪》中饰演的寇准也是身着红色学士衣夜审潘洪（八贤王做寇准的后盾，在潘洪傲慢且狡赖不招供时，二人定计，假设阴曹，才使潘洪吐露实情）。

（三）氅的文化内涵及其协调翻译

1. 开氅的文化内涵及其协调翻译

开氅是京剧中颇具权威的男性人物的常服，有时也作为礼服使用。开氅的"氅"字，本指一种鹙鸟的羽毛。古人用这种羽毛制成的裘衣即为"氅"。

由于开氅的样子庄重大方，因而它虽然不是官服，但在某些严肃的场合也可以作为礼服穿用。因为开氅需要外穿，所以可以将前襟解开，用左手将其拉向右侧。开氅的样式为大领（也称和尚领），斜襟右衽；长衣阔袖，袖端配有水袖。衣上的图纹多选用狮、虎、熊和象等图案。在衣服前后及领子的边缘部分多绣图案，形成一个宽边，而且底边尤宽。开氅的颜色也多种多样，有黑、白、红、黄和绿等，颜色各异。

---

❶ 阚艳华，董新颖，王娜，等. 中英文对照京剧服饰术语［M］. 覃爱东，陶西雷，校译. 北京：学苑出版社，2017：112.

❷ 经查证，将原文的"shuiliang"改为"Suiliang"。

❸ PAN XIAFENG. The Stagecraft of Peking Opera［M］. Beijing：New World Press，1995：119.

从穿用场合来看，开氅属于便服，适合高级武将、权臣在非礼仪场合（军旅或家居）穿用。因为其气派颇大，表现力强，也用于京剧戏剧中的某类主角人物，如占山为王的寨主、武艺高超的大将及侠士等。其形制为大襟和大领，氅长及足，带摆，有缘饰。

原文 The informal gown or Kaichang, as a garment for high ranking officials when they are in seclusion, is a full-length loose robe with a large collar and a lapel to go with a pair of wide sleeves dropping from somewhere around the ribs on both sides of the body. Ornaments，in needlework tinged with colors other than that of the robe, are trimmed all over the fringes and sleeve-cuffs in the shape of a band some 12 cm wide. Animal patterns are common. Often, a green gown has a lion pattern or a yellow one has a unicorn. In color, the gown has a somewhat broader range for it also appears in purple, blue or brown, in addition to the primary colors, i.e., red, green, yellow, white and black. As such, the lead in *Black Wind Scarf*, Gao Wang, is in an informal gown on his visit to a village, and Lian Po, one of the at leisure principle parts of *Reconciliation of the General and the Minister*, a historical piece portraying two statesman of the Zhao State, wears a similar gown to show he is at leisure. By virtue of its informality, however, this article was once shown to have new possibilities by the famous actor Hou Xirui when portraying Caocao, a third-century statesman, in *Battle at Wancheng*. Simply by dint of changing clothes on three separate occasions, he hinted at a legendary scandal of a love affair between Cao Cao and a charming lady in captivity, the local perfect's window. At first, Cao Cao substitutes his red officials robe for a purple informal gown in order to depict relaxation while the lady plays a zither. Then, he appears in a red informal robe accompanied by her to meet the request of her nephew for an interview. Finally, when a revolt id reported to him, he hastily rushes about dressed in a woman's red informal gown. Thus, the true story of the union by chance was more dramatically as well as romantically committed to the imagination of the audience than it would have been otherwise.❶

拙译 开氅是非正式礼服，属于宽松长袍、阔袖，为高官隐居时穿用。袍服边缘和袖口上缝有宽度约为12厘米的饰带。开氅通常都饰有动物图案，如绿色长袍饰有狮子图案或黄色麒麟。开氅的颜色范围也很宽泛，除了原色，即红色、绿色、黄色、白色和黑色外，还有紫色、蓝色或棕色。因此，《黑风帕》中的领袖高旺，走访各村时就穿着这种非正式长袍；《将相和》中的廉颇也穿着类似的袍服以体现其休闲的状态，这两位都是历史上国君的代表人物。然而，正因为开氅的非正式性，知名演员侯喜瑞在《战宛城》中扮演三国时期的曹操时，就选用了开氅展示曹操的新形象。京剧中仅仅通过三次改变曹操的着装方式，就暗示了曹操与传说中的一位当地迷人女子之间的爱情

---

❶ PAN XIAFENG. The Stagecraft of Peking Opera［M］. Beijing: New World Press, 1995：117–118.

京剧服饰大衣箱文化阐释及翻译协调美

第三章

73

故事。故事开始，曹操身着紫色开氅而非红色官袍，体现其听女子弹古筝时的放松心态。随后，曹操身着红色开氅陪同女子面见她侄子。最后，当有人向曹操报告有叛军时，曹操匆忙穿上女式的红色开氅。这出三国故事因为剧情中的诸多服饰上的巧合更具戏剧化，充满了浪漫气息，也更加吸引观众。

原文 Warriors wear Kaichang, which is a kind of open cloak. Typically, it resembles the outer vestment worn by a Buddhist monk or Taoist priest, and it has water sleeves attached to the cuffs. The front buttoning are on the right and embroidered with lions, tigers, or other animals, never dragons. The left and right sides of the cloak are fitted with protective armor.[1]

拙译 开氅主要为武将穿着，式样为斜大襟、和尚领，腋下开衩有硬摆，长袖带水袖，长至足面，前后襟绣有狮子、老虎和其他猛兽图案，但永远不能出现龙的图案。开氅的左右两侧有防护装置。

原文 The Kaichang is a traditional Jingju-developed garment that blends the cut of the Mang and the Xuezi. The Kaichang is worn for informal or private occasions by chancellors, advisors, military strategists. It has a long rectangular sleeve with water sleeves. The front neckline with a straight crossover closing to the right finished with a wide straight collar band. The sides and back are taken from the mang cut; the fronts lap over the back at the side seams with extensions tied to the center back of the neck. The color is made in upper five colors. The ornamentation on the surface are large powerful animals, such as lions, tigers, and elephants.[2]

拙译 开氅融合了蟒和褶子的裁剪方法。开氅是大臣、顾问和军事参谋在非正式或私人场合穿用的袍服，斜襟右衽，样式为大领（也称和尚领），长衣阔袖，背面和侧面都仿效蟒袍的裁剪方式，前片在侧面盖过后片，延伸部分与脖颈后面的中间位置相接。开氅的颜色属于上五色，其装饰图案多为猛兽，如狮、虎和大象。

2. 鹤氅的文化内涵及其协调翻译

鹤氅是军师或仙翁的服装。样式在整体结构上与开氅有区别，如鹤氅背后无摆，大领，袖口均绣有边子，有腰梁飘带。周身彩绣八个团鹤。色彩上通常有黑、天青、蓝、紫。目前此服装在舞台上使用很少。不妨对照下列英汉表达加深对其的了解。

原文 Somewhat similar to the eight-diagram garment is the so-called "crane cloak" (Hechang), which, with designs of rounded cranes instead of both the cosmos diagram and the

❶ 赵少华. 中国京剧服饰［M］. 北京：五洲传播出版社，2004：38.

❷ BONDS ALEXANDRA B. Beijing Opera Costumes: the Visual Communication of Character and Culture［M］. Honolulu: University of Hawaii Press，2008：145.

eight diagrams, is worn by such geniuses as listed above; for example, Xu Shu in *Recommending Zhuge Liang Before Departure* wears a green cloak with white crane designs when he puts in a good word for his friend Zhuge Liang.❶

拙译 在京剧服饰中与八卦服相似的是八仙穿着的鹤斗篷（鹤氅），其四周装饰有仙鹤，而非波斯菊和八卦图。例如，在《荐诸葛》中徐庶身着绣有白色仙鹤图案的绿色斗篷，推荐朋友诸葛亮给刘备。

3. 团花开氅的文化内涵及其协调翻译

团花开氅用于高级文官，其形制为斜领，大襟，衣长及足，后身有摆。刺绣方式为绒绣或平金绣，衣边及袖口绣缘饰纹样，不单镶边。主要纹样为团花（周身彩绣共10个团花）、吉祥八宝、如意等图案。

原文 团花开氅用于高级文官，斜领，大襟，衣长及足，后身有摆。周身彩绣团花、吉祥八宝、如意等图案。❷

译文 Tuanhua Kaichang (Secondary Ceremonial Robe Embroidered with Rounded Patterns) is used by senior civil officials, with a crossover fastening, and it features buttons on the right, extends down to the feet, and has a back hem. It is embroidered with rounded patterns of flowers (10 flowers in all), the eight auspicious treasures or "Ruyi".❷

4. 麒麟开氅的文化内涵及其协调翻译

麒麟开氅主要为武将、侯爵生活中的便装。其形制为斜领，大襟，衣长及足，后身有摆。选取传说中的神兽麒麟为开氅的主要纹样，插底为博古图案（又称"四艺"吉祥），象征人物位高权重。麒麟开氅一般为黄色，绒绣；若用黑色，则采用平金绣法；有时也用紫红色。

原文 麒麟开氅，斜领，大襟，衣长及足，后身有摆，主要为武将、侯爵生活中的便装。麒麟为传说中的神兽，为开氅的主要纹样，插底为博古图案，象征人物位高权重。服色一般为黄色，有的也用黑色或紫红色。❸

译文 Qilin Kaichang (Kirin Open Robe) is an informal robe for an Official, with a crossover fastening, and it features buttons on the right, extends down to the feet, and decorated with a Kirin pattern. It is for both private and informal wear, and is worn by high-

---

❶ PAN XIAFENG. The Stagecraft of Peking Opera［M］. Beijing: New World Press，1995：118–119.
❷ 阙艳华，董新颖，王娜，等. 中英文对照京剧服饰术语［M］. 覃爱东，陶西雷，校译. 北京：学苑出版社，2017：17，略有改动.
❸ 同❷16.

ranking military and government officials. The robe can be yellow (cross-stitch with yarn), black (golden couching stitch), or purple in colour.❶

（四）帔的文化内涵及其协调翻译

帔，源于宋代贵族妇女的大袖褙子礼服，原本窄袖，对襟，大领，直贯底摆。后至明末，其袖式逐渐演变为大袖，领式也由长大领缩为半长大领。

帔旧称披风。长领、对襟、大袖（带水袖），左右胯下开衩。男帔长及足，女帔稍短，长仅及膝。花色极多，周身为平金或绒线刺绣图案纹样。一般作为中级官吏、豪宦乡绅及其眷属在家居场合通用的常服和便服。剧中表现夫妻关系时，多穿色彩和纹样上完全相配一致的帔，所以称为对儿帔，具有舞台画面的整齐美。

另有观音帔，为观世音菩萨专用。观音帔为白色，圈银绒绣绿竹或黑竹，还有的平银绣竹。观音帔专用的竹叶纹样具有象征性，表现人物所处的自然环境——南海仙山上的竹林。

帔的服装造型虽属于中国古代传统服饰的主流（宽袍大袖式），但它显然有别于"不壮不丽无以重威"的礼服——蟒袍。比起蟒来，它突破了"全封闭式"的服装造型，以对襟体现自由开合的宽松感，以向下的两条垂直线给人以流畅修长的美感。帔的另一个特点是：男团花帔和女团花帔一般是成对的夫妇团花帔，用于中级官吏、豪宦乡绅及其眷属。通常选用较沉着的色彩（紫红、古铜、深蓝、秋香色等），多绣团寿字、团鹤等象征福寿延年的吉祥图案。因女红帔与男红帔配对使用，二者色彩、纹样完全一致。女红帔下身系红色绣花大褶裙。

帔（Pei coat / everyday dress）与蟒、褶等都属于京剧戏箱中不可缺少的行头，既可以为帝王、权贵、中青年官员、在职与告老还乡的老年高官穿用，也可以为员外乡绅一类的人物当作便服。皇帝除去在上朝等重要场合穿蟒袍外，在其他办公、设宴等场合一般都穿带有便服性质的帔。皇帝所穿帔也叫为黄帔，明黄色，绒绣团龙，对襟的开合实际上是使用暗纽襻（宝剑头式飘带一般不打结），同时男用帔需内衬褶子。皇帝用团龙，皇后、贵妃用团凤，太后用团龙凤。除此之外，视人物年龄身份，或用团花、团寿字，或用枝子花。皇室成员用黄色，状元登科或新婚典礼、喜庆团圆场合的夫妇用红色，老年人用秋香色，其余人物的用色没有严格界定。员外所穿帔也称为员外帔。

帔一般由缎料制作，大领、对襟、长衣、阔袖，袖端缝有水袖，前胸正中对襟处有一对飘带，左右两侧自胯部向下开衩。该行头的穿用者多具有较高的社会或经济地位，同时该行头又属于便服，融庄重感与休闲性于一体。特别是帔的前身正中飘带以下，以及左右两侧均是分开而不闭合的，因而在人物行走、转身及做各种动作之际往

❶ 阙艳华，董新颖，王娜，等. 中英文对照京剧服饰术语［M］. 覃爱东，陶西雷，校译. 北京：学苑出版社，2017：16，略有改动.

往衣襟飘拂，洒脱自如，别有韵致。同时也正因为帔具有衣摆纷披、易于将双腿显露于外的特点，所以穿帔时都要在里面穿一件衬褶子。

综上所述，帔的意义不仅在于表现某种实用功能，还在于显示人物的身份、尊贵和富有。因此帔上常绣团龙、团花、仙鹤和寿字等图纹，显露着人物的身份。据《菊部丛刊》所载的《不可不知录》❶中说，帔的色彩原来只有红、黄、蓝三种，后来才出现其他颜色，在下文的英汉表达中同样可见一斑。

原文 The informal robe (Pei) falls into two main types: the male's, which is rather long; and the female's, which is comparatively short. As a piece of everyday wear, both man and wife may wear suitable ones at the same time on the stage. This is what the characters Liu Yanchang and Wang Guiying in *Magic Lotus Lantern* and Meng Mingshi and his wife in *Pavilion of the Imperial Tablet* wear. Thus , the informal robe has gradually come to be tagged a couple's robe (Dui'erpei). In form, it has a double lapel and a slit from under the hip down to the feet on both sides. Its overall embroidery varies with the dramatic personae as to what roles they play: an emperor uses a yellow informal robe in dragon-patterned embroidery, while an empress, a dowager empress and most of the ladies-in-waiting may each have a phoenix-decorated one. High officials, in contrast, usually dress in blue, black, purple or brown informal robes with decorations such as a white crane, a green deer, or a stylized Chinese character signifying "longevity". One style of these robes is red and decorated with annular embroidered flowers and is regularly used as a weeding robe by dignitaries.❷

拙译 非正式长袍（帔）主要分为两类：一类是属于男性的相当长的帔；另一类是女帔，则相对较短。作为日常穿用的袍服，男子和妻子在舞台上可以同时穿用。正如《宝莲灯》中的人物刘彦昌和王桂英，以及《御碑亭》中的孟明时及其夫人，在剧中都身着帔。于是，非正式的长袍——帔，就逐渐被贴上了夫妻长袍的标签（对儿帔）。在形式上，对襟处的左右两侧自胯部向下开衩，其整体刺绣因戏剧人物角色而变化。皇帝身着非正式黄色刺绣龙袍，而皇后、皇太后和大多数的答应每人都有一件凤凰图案的霞帔。相比之下，高官通常穿着配有白鹤、绿鹿或寿字图纹的非正式长袍，颜色为蓝、黑、紫或棕色。这些长袍中有着环纹图案的红色长袍一般用作达官显贵的婚服。

原文 Everyday dress worn by members of the imperial family, nobility and high officials is Pei. A Pei is a satin jacket with buttons down the front. It has a long collar and loose "water" sleeves attached to the cuffs. Pei is long enough to cover the insteps. Pei jackets are embroidered

❶ 周剑云. 菊部丛刊［M］. 上海：上海交通图书馆，1918：13.

❷ PAN XIAFENG. The Stagecraft of Peking Opera［M］. Beijing：New World Press，1995：113–114.

with dragons, phoenix, flowers and birds. When a female character wears a Pei, she also wears a skirt.❶

拙译 帔是身份尊贵者穿着的居家便服。样式为对襟长领，宽袖带水袖，长及足面，缎子上绣花鸟走兽等图案。女帔大致样式与男帔相同，但女帔长度过膝，穿着时需要穿裙子。

原文 "Pei" is kind of robes for Emperors, Queens, Officials, landlords and their families at the casual occasions. "Pei" also has two styles for male and female. From color standpoint, the bright yellow color is tailor-made for the emperors and queens. the robes for emperors are embroidered with dragon, and the robes for queens are embroidered with phoenix. When "Pei", the robes are used for a couple, the colors and patterns are exactly the same. So this kind of robes are called as "Couple Robes". The red "Couple Robes" are specially for the newlywed. The round flower pattern is embroidered for married women and a piece of vivid flower is embroidered for young girls.❷

拙译 帔是帝后、官员等身份尊贵者和乡绅及其家眷在家居场合通用的常服，分为男帔和女帔两种。在颜色的使用上，明黄色为皇室专用，皇帝穿用的帔上绣龙纹，皇后的帔则绣凤纹。一般当夫妇二人同时穿帔时，男女帔的色彩、纹样完全一致，故称为对儿帔，其中红色的对儿帔尤其适用于新婚夫妇。已婚的女性角色一般穿绣团花的帔，而未婚的少女穿绣枝子花的帔。

原文 Pi evolved from the Beizi which during the Song dynasty. The Beizi was worn for nonofficial and informal events. It is considered more formal than Xuezi but less important than the Mang. It is commonly worn for daytime wear at noncourt functions. It opened down the center front, with a wide band going around the neck and then stopping at mid-chest. Cut without a shoulder seam, side-vented, and the the straight sleeves are longer than the wrists with water-sleeves. The men's Pi is full length, the women's is only knee length. The color range follows the standard convention. The ten dragon round designs found on the Mang are also used on the Pi, some geometric designs and flowers also on the surface of Pi.

Nüpi shares the characteristics of the Pi, also without a shoulder seam and reaching to the knee, and long straight sleeves. Young ladies crepe Nüpi and mature wear a Nüpi of satin. The empress's Nüpi has ten rounded phoenix designs on it, the symbol for high-ranking imperial women. Young, unmarried women wear a Pi with freely arranged floral fragments.❸

---

❶ 赵少华. 中国京剧服饰 [M]. 北京：五洲传播出版社，2004：30.

❷ 谭元杰. 中国京剧服装图谱 [M]. 北京：北京工艺美术出版社，2008：53.

❸ BONDS ALEXANDRA B. Beijing Opera Costumes: the Visual Communication of Character and Culture [M]. Honolulu：University of Hawaii Press，2008：124.

帔是由宋代的褙子演变而来。褙子当时在非官方或非正式场合穿用，比褶子更正式，但是没有蟒那么重要。帔通常在白天的非宫廷场合穿用。前开襟，脖子四周镶有宽边儿，并垂至胸前的中间位置。裁剪上不设肩缝，侧开口，直袖长过于腕并缝接着水袖。男款帔全身长，女款帔长及膝盖。帔的颜色范围遵循惯例。蟒袍上十条龙的图案也同样适用于帔，同时还有一些几何图案和花卉图案。

女帔具备帔的特点，没有肩缝，长及膝盖，且长袖。从材质上看，年轻女性穿绉绸，而成熟女性则穿大缎。皇后的女帔绣有十个团凤图案，是皇室女性最高级别的象征。年轻未婚女性的帔一般绣有枝子花纹样。

译文剖析 帔作为居家式穿着的服饰，来自宋朝的褙子。对这一文化背景和内涵，赵译将其表述为 "a satin jacket for everyday dress"（日常穿用的服装）；谭译表述为 "a kind of robes wear at the casual occasions"（在休闲场合穿着的袍服）；亚历山德拉的译文表述为 "worn for nonofficial and informal events, worn for daytime at non-court functions"。对比剖析，笔者认为赵译的表述 "everyday dress" 属于日常的服装，可以理解为非礼仪性的服装，但不等于居家场合穿着的服装，并不能体现其穿用的特定场合。亚历山德拉译文解释的 "non-court" 非宫廷场合，似乎有点过于绝对。谭译的解释 "casual occasions" 休闲场合，比较符合原文内容，也不会让读者产生歧义。另外，对于 "帔" 本身的解释，赵译的表述 "a satin jacket" 更为客观，比较符合现代人的思想方式。"jacket" 夹克，即一种休闲的对襟式外套，这和 "帔" 的定位巧妙地吻合。

对于帔的样式 "对襟长领，衣长及足，阔袖带水袖"，赵的解释是 "buttons down the front, and a long collar, water sleeves, long to cover to the insteps"，亚历山德拉译文的表述为 "It opened down the center front, with a wide band going around the neck and then stopping at mid-chest. Cut without a shoulder seam, side-vented, and the the straight sleeves are longer than the wrists with water-sleeves"。相比之下，亚历山德拉将帔式样上的各个细节都淋漓尽致地给以阐释，包括对襟、对襟的飘带、左右两侧分开并开衩、阔袖有水袖等。笔者认为可以在此基础上融合赵译对于 "长领和长至脚面" 的阐释，特别是 "long to cover to the insteps"，十分巧妙与灵活。"insteps" 在《牛津高阶英汉双解词典》中的解释为 "the top part of the foot between the ankle and toes"[1]，在脚踝和脚趾中间，即脚面的位置。这个单词的选取与协调翻译更加符合 "长至及足" 的本义。亚历山德拉对帔款式上 "两侧开衩" 的翻译也较为忠实，"vent" 在《牛津高阶英汉双解词典》中的解释为 "a thin opening at the bottom of the back or side of a jacket or coat"[2]，衣服后边或侧边的开衩，而将 "vent" 转换为 "side-vented" 较符合 "两侧开衩" 的含义。

对于帔颜色、纹样和女帔的文化内涵表达，赵少华、谭元杰和亚历山德拉皆对此做出了详细阐述，包括其颜色的多变、图案的丰富，以及女帔在式样上与帔的一致等，

---

[1] 霍恩比. 牛津高阶英汉双解词典［M］. 王玉章，赵翠莲，邹晓玲，等译. 7版. 北京：商务印书馆，2010：1058.
[2] 同[2]2236.

以及纹样和颜色有女性专属的牡丹、凤凰和粉色等细节。在介绍女帔的时候，均提到了其通常和帔形成一对，即夫妻形式的对儿帔。

1. 男帔的文化内涵及其协调翻译

男帔，对襟，长领，阔袖带有水袖，衣长及足，左右开衩。周身彩绣草龙、仙鹤、蝙蝠和团花等图案。服色有黄、红、蓝和秋香色。主要为帝王、官宦乡绅及其家眷在居家场合时穿用。阙艳华等对此给予如下英文阐释：

Nanpi (Male Robe) is a satin jacket with buttons down the front, left and right thigh slits, a long collar and loose "water" sleeves attached to the cuffs. The entire robe extends down to the feet. It is embroidered with dragons, cranes, bats and rounded patterns of flowers in the modest colors such as yellow, red, blue and chartreuse, which is worn by emperors, officials, squires and their relatives, at home or on informal occasions.[1]

2. 女帔的文化内涵及其协调翻译

样式与男帔基本相同，但尺寸略短，衣长及膝。周身彩绣仙鹤、蝙蝠、团凤、团花等图案。服色有黄、红、蓝、秋香色等。主要为皇后、贵妃、贵妇穿用。阙艳华等学者对此给予如下英文阐释：

Nüpi (Female Robe）is similar Nanpi in shape, but it has a lightly shorter length, and only reaches the knees. It is embroidered with cranes, bats, rounded patterns of phoenixes or flowers and appears in modest colors such as yellow, red, blue and chartreuse. It is worn by empresses, imperial concubines and noblewoman.[2]

对比上述男帔和女帔的汉语表达与英文阐释，都较好地传递了其文化内涵，方便西方读者理解。

（五）八卦衣的文化内涵及其协调翻译

八卦衣专用于足智多谋，且有道术的军师或谋士类人物，其基本样式与开氅相似，上绣八卦及太极图，斜襟大领，阔袖带水袖，四周有花边，身后无摆，腰部略向里收，腰际亦有花边，并垂飘带两条。周身绣着道教纹样，有八卦和阴阳鱼组成的太极图；布局十分严谨，构成了一幅完整的太极图。在服装的中心位置还绣有太极，以黑为阴，以白为阳，阴阳旋转于圆形之中。在太极图案的周围，八卦符号排列有序。左肩绣的符号象征天（乾），右肩绣的图案象征地（坤）。前身左下部的图案象征风（巽），前身

---

❶ 阙艳华，董新颖，王娜，等. 中英文对照京剧服饰术语［M］. 覃爱东，陶西雷，校译. 北京：学苑出版社，2017：18，略有改动.

❷ 同❶19，略有改动.

右下部的图案象征雷（震），后身左下部的图案象征山（艮），后身右下部的图案象征沼泽（兑）。左袖外侧图案象征水（坎），右袖外侧图案象征火（离），八卦字在服装上的具体含义是"肩担天地，胸怀风雷，背负山川，袖藏水火"。这种抽象符号作装饰纹样只专用于塑造知天文地理的智慧人物，在京剧服装中较为少用。衣色一般为青色或天青色、宝蓝色。笔者摘录了中外三位学者对于八卦衣的介绍，并采用英汉对照的方式，方便读者多角度理解其文化内涵。

原文 An eight-diagram garment or Baguayi specially identifies strategists, such as Zhuge Liang (181-234) in *Empty City Ruse*, Liu Bowen (1311-1375) in *At the Harbor of Jiujiang* and the fictitious character Wu Yong in *Huangni Ridge,* although the latter two each have a silk braid bound to the garment and hold a whisk. This costume, in black and purple only, is notable particular for its all-over designs of the eight diagrams in golden thread and for both front and back cosmos diagrams beneath a larger collar. Two streamers swing loosely in front from the waist.❶

拙译 八卦衣用于类似《空城计》中足智多谋的诸葛亮（181—234）、《九江口》中的军师刘伯温（1311—1375），以及《黄泥岗》中的传奇人物吴用等类似人物。尽管后面两位的袍服上都有丝带，并手持拂尘，但是这类袍服一般只有黑色和紫色，全金属线绣制的八个图形以及大领子下正面和背面的波斯菊图案非常引人注目，腰际的两条飘带垂落，给人以飘逸自然之美。

原文 The Baguayi is usually maroon, dark blue or the deep Qing color. Any of these colors can be worn in a given scene. The distinctive feature of the Baguayi is the pattern of the eight trigrams arranged around the surface, along with the Yin and Yang symbol that appears on the front and back chest.

The eight trigrams are thought to have come from the markings of a tortoise shell that were then developed by the legendary emperor Fuxi (2852 bc) into mystic symbols. The symbols are made from the eight possible patterns that can be derived by combining a continuous line, the yang principle, with a broken line of the yin principle, in sets of three.

Each symbol contains a different meaning, and they are typically arranged on the garment for maximum effect. The sky (Qian) is on the left shoulder, and the earth (Kun) is on the right. On the lower left front is the symbol of the wind (Xun), while opposite it on the lower right is thunder (Zhen). On the back, the mountains (Gen) are located on the lower left, and lakes (Dui) are on the lower right. The outside of the left sleeve is decorated with the sign for water (Kan), and fire (Li) appears in the same location on the right. The overall meaning of the eight trigram emblems is "to shoulder the sky and earth, to have wind and thunder in the heart, to rely on the mountains and

❶ PAN XIAFENG. The Stagecraft of Peking Opera［M］. Beijing：New World Press，1995：118.

京剧服饰大衣箱文化阐释及翻译协调美 ○ 第三章

rivers on the back, and to store water and fire in the sleeves".

The trigrams are couched in gold or platinum, and in some versions of this costume they are surrounded by a circular floral design. The Yin and Yang symbols on the chest and back represent the positive and negative forces in the universe. The dark yin symbolizes the earth, moon, darkness, and feminine aspects, while the light yang represents heaven, sun, light and maleness. The combined symbol is usually couched in two colors of metallic thread. The combination of the trigrams and Yin and Yang represent the wearer's knowledge in astronomy and geomancy, religion and military strategy.[1]

拙译　八卦衣通常是栗色、深蓝色或深青色，这些颜色都可以在特定的场景中穿用。其特色是排列着八卦图案，前胸、后背都有阴阳符号。

八卦起源于远古时期，发现于白龟甲壳上，后由传说中的伏羲氏（公元前2852年）发展为神秘符号。这些符号由代表八种可能性的图案构成，连续线为阳爻，间断线为阴爻，三个为一组。

每个符号包含不同含义，排列在服装上一起产生最大影响。天（乾）在左肩，地（坤）在右肩；左下方是风（巽）的象征，右下方是雷（震）的象征；在背面，山（艮）位于左下方，湖（兑）位于右下方；左袖的外侧饰有水（坎）标志，火（离）出现在右侧的相同位置。八卦象征的整体意义是"肩负天地，心中有风雷，依靠背山、河流，水、火在袖中存"。

八卦图案采用金或铂线绣制，有些改良版中的八卦图是借用花艺设计来展示的。胸部和背部的阴、阳符号分别代表宇宙中的正面和负面力量。阴象征着地、月和黑暗，以及女性特征；明亮的阳则代表天、日、光和男性的阳刚。阴阳符号组合通常借用两种颜色的金属线绣制而成。阴阳结合的卦相体现了穿着者精通天文、风水、宗教和军事策略的特征。

原文　The eight diagrams are eight combinations of three whole or broken lines, formerly used in divination. Thus eight-diagram cloaks are for semi-human, semi-divine characters who often serve as military advisers.[2]

拙译　八卦衣用于军师等足智多谋的人物，其形制与开氅相似：斜襟大领，阔袖带水袖，衣身及袖口镶边。八卦衣不同于开氅之处在于，其身后无摆，腰部向内收，腰梁下垂有两条带金绣的青色飘带（八卦衣最显著的标志在于其带有象征意义的太极图装饰纹样。太极图在八卦衣上代表："肩坦天地，胸怀风雷，背负山川，袖藏水火"）。

❶ BONDS ALEXANDRA B. Beijing Opera Costumes: the Visual Communication of Character and Culture［M］. Honolulu: University of Hawaii Press，2008：174.

❷ 赵少华. 中国京剧服饰［M］. 北京：五洲传播出版社，2004：64.

（六）法衣的文化内涵及其协调翻译

法衣为有法术的道士、神仙所穿的服装。法衣的样式为大领，对襟，无袖如斗篷，而左右开有袖口，双手及法衣里面所穿褶子的水袖可以从里展露于外，前后身都近于方形，衣上绣制宽边，前身正对襟处饰有长飘带。法衣上可绣八卦、太极、日月、云朵、仙鹤、行龙等图案。法衣的颜色有红、黄、绿、白、黑等几种。法衣在所有京剧行头中，样式奇特，独树一帜，常用来满足相对应的神话传奇性故事情节之需要。笔者摘选有关法衣的两个不同表述，英汉对照，方便西方读者理解其文化内涵。

原文 This Fayi has the eight trigrams symbols embroidered on the border. The Fayi is worn by both priests and immortals, including Zhuge Liang, the Daoist military strategist in *the Three Kingdoms Stories* in *The Gathering of Heroes* (*Qunying hui*). ❶

拙译 法衣上绣有标志性的八卦图，为道士和神仙穿用，如《三国演义》之《群英会》一折戏中的道家和军事家诸葛亮，就穿着八卦衣。

原文 When worn by Zhuge Liang, the Fayi is blue gray with dark blue borders. The Fayi can be decorated with the eight Daoist attributes or the eight trigrams.

Underneath the Fayi, the actor will wear high-soled, high-topped boots, black trousers and a black Xuezi or Baguayi robe.

The Fayi can be worn with the Lianhua Guan (lotus crown) with religious symbols. Zhuge Liang has a hairpiece of long hair (Pengtou) worn loose on the back of his head when he is wearing this headdress. Hair spread in the back (Pifa) is a Daoist characteristic. ❷

拙译 诸葛亮的八卦衣是蓝灰色的，并镶有深蓝边儿。法衣上装饰有道家八卦图。在法衣的下面，演员穿着高底厚靴、黑裤子、黑褶子或八卦袍服。

法衣还常和带有宗教意味的莲花冠搭配在一起。饰演诸葛亮、穿戴这种头饰时，还会在后脑勺上佩戴名叫"蓬头"的假发头套，松散着几绺长发。道教的特点是头发散于肩部（披发）。

（七）僧衣的文化内涵及其协调翻译

僧衣斜襟开身，但袖宽相当于身长，无水袖。有黄、紫、香色和灰色等。美国学者亚历山德拉对此有着如下详细的描述。

❶ BONDS ALEXANDRA B. Beijing Opera Costumes: the Visual Communication of Character and Culture [M]. Honolulu：University of Hawaii Press，2008：175，略有改动。

❷ 同❶175.

原文 Monks appear regularly in traditional Jingju stories, and they are dressed in garments similar to those worn by their real-life counterparts. The long Sengyi has a straight crossover closing to the right, with a wide collar band similar to the Xuezi and the garment worn by actual Buddhist monks. The sleeves are wider than those of the Xuezi, and there are no water sleeves. Usage. An abbot in a Buddhist monastery wears this style of robe.

This robe comes in colors worn in Buddhist monasteries: gray and shades of light brown, beige and yellow, with the yellow being worn by the head monk.

Characters wearing this costume carry a small string of brown wooden prayer beads in their hand and wear flat-soled shoes on their feet.

The Sengmao (monk hat) is peaked in front and has upturned flaps above the ear area. The character for Buddha is embroidered above the forehead.[1]

拙译 在传统的京剧故事中，经常有僧侣角色，其穿着类似于现实生活中僧侣的装束。长款的僧衣右侧用宽大的领带打结儿，类似于褶子和真正的佛教僧侣穿用的服装。僧衣的袖子比褶子的要宽，并且没有水袖。佛教寺院的住持常穿这种风格的长袍。

这件僧衣长袍的颜色是佛教寺院的颜色：灰色和浅棕色，米黄色和黄色。其中黄色通常由住持穿着。

穿僧服的人还通常手持棕色木制念珠，脚踏平底鞋。

僧帽（和尚帽子）一般前部高耸、耳部下垂，佛教信徒僧帽的额部上方位置还绣有佛像。

原文 Xiaosengyi (Young monk's robe) for young monks are calf length and have standard-width sleeves, though some examples have the wider sleeve of the Sengyi. These robes have no water sleeves. The collar band is wide and straight, closing diagonally to the right.

This style of robe is worn by lower-ranking and younger monks, played by either Chou or Xiaosheng.

The robe is usually gray and may have a black collar band.

A long string of wooden beads is worn with this garment. Flat-soled shoes in gray or black and heavy white cotton socks cover the lower legs and feet. Black trousers are tucked into the tops of the socks.[2]

拙译 小僧衣（小和尚的长袍）是年轻僧侣穿的长袍，长及小腿。尽管有时其袖口很宽，但实际袖子的宽度也有标准。领口宽且直，右交领。

---

❶ BONDS ALEXANDRA B. Beijing Opera Costumes: the Visual Communication of Character and Culture〔M〕. Honolulu: University of Hawaii Press，2008：175.

❷ 同❶175-176.

这种款式的袍服一般是由地位低下或年轻的僧人穿用，如京剧中的丑角和小生。

僧衣通常是灰色，有时配有黑色领口。

穿僧衣时要佩戴木质念珠搭配，穿灰色或黑色的平底鞋，并用厚白布袜裹住小腿和脚，黑裤子也要塞进袜子里。

原文 Usage. The actual monk's cloak is worn with the five-paneled hat described below for reading the rites in Buddhist ceremonies. The name of the garment implies the situation in which it is worn. In traditional Jingju, however, the garment has been transformed into status dress and represents who the monk is rather than the situation. [1]

拙译 在佛教僧侣诵经仪式中，实际穿用僧袍时需佩戴下面即将提到的佛教场合镶嵌有五尊佛像的帽子，袍服名字本身就暗示了其寓意。在传统的京剧中，僧衣已经演化成为地位的象征，即代表了谁是和尚，而不是指具体的场景。

原文 Color and ornamentation. The head monk cloak comes in red only, with a couched linear pattern resembling bricks to provide symbolic protection from demons and ghosts. In the past, there were ninety-nine bricks, with eighty-one lines between them, a numerical combination relating to the Chinese belief in nine as an auspicious number. [1]

拙译 在颜色与配饰上，住持僧人的外袍是红色的，上有类似砖块的长方形图案，象征性地保护人免受恶魔和鬼魂的伤害。过去的住持僧袍上有九十九块砖形图案，其间有八十一条线，这与中国人相信九是吉祥数字相关。

原文 Headdress. The Wufoguan (five-Buddha headdress) is flared, has five Buddha figures on the front panels, and white streamers that hang over each ear.

Sengkanjian or Wusengyi (Military monk vest): A military monk wears a black Kuaiyi under a distinctive black Sengkanjian, with two moving-water layers of pleated fabric in red-orange and yellow attached to the hem. The neckline of the vest falls straight from the shoulder and is decorated with a contrasting band that ends with the fungus-shaped. [1]

拙译 在头饰方面，五佛冠（五佛头饰）呈喇叭形，前面镶嵌有五尊佛像，耳垂白色飘带。

而僧坎肩或武僧衣（武僧马甲）指武僧在黑色快衣上外套特有的黑色僧坎肩，下摆处配有行云流水般的橙红色和黄色包边褶裥。坎肩领口处到肩部镶有对比色边带，以蘑菇形收边。

---

❶ BONDS ALEXANDRA B. Beijing Opera Costumes: the Visual Communication of Character and Culture [M]. Honolulu: University of Hawaii Press, 2008: 177.

（八）褶子的文化内涵及其协调翻译

褶子又称道袍，源自明朝斜领大袖衫，其用途最广且装扮形式最多，是最为常见的袍服类服装。褶子是一种斜领长衫式样的便服，大领右襟，长袍阔袖，袖端有水袖。褶子在京剧中也是一种通用性最强的便服，无论是皇帝、高官还是平民百姓，无论是穷人还是富人都可以穿着。

褶：“袷也”❶，“袷”即“夹”的意思。古代实际生活中的褶子就是一种夹衣服。这种夹衣既可以是布衣，也可以是锦衣华装。京剧中的褶子样式为大领、右大襟，阔袖且袖端加水袖，不挂摆，是与蟒和官衣一样的长袍。因使用的衣料不同而有软、硬之分，大缎质地的褶子为硬褶子，绉绸质地的为软褶子；根据是否有刺绣装饰，又有花、素的区别，无刺绣的为素褶子，有刺绣的为花褶子。褶子上所绣花纹则是多种多样的，或花卉或禽鸟或其他吉祥图案等。这些纹饰的繁简与布局也有许多种类，有的仅在底摆衣角和领下有少许绣饰，有的则通身及双袖遍布绣饰，有的绣以团花，有的则绣以散点式的碎花，有的外绣里不绣，有的则连里子也绣上一些较为简略的纹饰。褶作为一种多功能服装，可以单独穿着又可以与帔、箭衣等服装搭配。

在褶子的穿法上，因剧情和人物的不同，既可在右侧腋下将衣襟系好，也可不系衣襟而用左手将大襟拉向左侧，表示所穿褶子为外衣。有的人物穿裤子不系衣带，也有的人物穿褶子时须在衣外系大带或丝绦。有的京剧行头只用于特定的人物，因而其色彩也有一定限制，如绿色多用于武职；而官衣为文官官服，故官衣类行头中并无绿官衣。但褶子由于极富通用性，为了与各类人物及多种情节相适应，其色彩也格外丰富，众彩俱备。

褶类的品种很多，按纹样和形制区分，共有21个品种。男式褶源于明代男子便服——斜领大袖衫，这是中国古代传统的便服形制，远在唐宋时代，平民及士人都可穿用。男褶身长及足，腋下开衩。通常在软或硬的绉缎上绣具有象征性的独特纹样，如书生秀才用角隅纹样、将领用二方连续纹样、衙内角色用散点式小碎花纹、英雄和侠士也用散点布局等。女式褶则是在明代“小立领对襟窄袖袄”的基础上发展而来。女褶子的式样为对襟立领，长过膝，开衩，袖端有水袖。小立领这一领式是程朱理学的思想禁锢在妇女服饰上的体现（妇女脖颈要用立领加以遮盖）。

褶子分花、素两大类。花褶子中又有武生褶子与小生褶子之分。武生褶子不但外面绣有鸟兽花卉，里面也有花绣，可以敞着穿。小生褶子过去也是满身花绣，近时多用折枝花缀其一角，显得淡雅、潇洒。素褶子多用于平民百姓，其名目有多种：①青褶子，即黑色素褶子，有白领，为社会地位较低者所穿。②穷衣，在青褶子上补缀若干块不规则的杂色绸子，为穷途潦倒的书生所穿，因其日后富贵，故又称此衣为“富贵衣”。

---

❶ 陈彭年. 宋本广韵［M/OL］.［2019-01-17］. https://ctext.org/dictionary.pl?if=gb&char=%E8%A4%B6&remap=gb

③海青，青褶子但用青领者，为家院一类奴仆所穿，故又称院子衣。④青袍，布制的青褶子，既无白领，又无水袖，为站堂的衙役所穿。⑤紫花老斗衣，布制的土黄色素褶子，为农民或其他劳动人民所穿。⑥短跳，即短式的紫花老斗衣，用时腰束白布短裙，亦为劳动人民所穿。⑦安安衣，短式的布制蓝色素褶子，为平民家的儿童所穿。此外，旦角穿的褶子比男褶子稍短，亦有花、素之分。老旦穿的布制素褶子多用青领。另有一种对襟的黑色素褶子，滚蓝边，称青衣，为中青年贫妇所穿。褶子又称道袍，僧、道也可穿用。素褶子还可以作为蟒、官衣、开氅、帔、花褶子等的衬衣用。

文小生花褶用于书生、秀才。绒绣枝子花（角隅纹样），纹样简洁鲜明，布局均衡，色彩淡雅明快，服色十分突出，给人以清秀洒脱之感，可以衬托出人物文静风流的性格和气质。武小生花褶用于身份高的青年儒将，绒绣，缘饰为二方连续纹样。花脸花褶用于具有粗犷豪放性格的人物类型，一般绒绣大枝子角花，或平金绣散点式大流云，敞穿。文丑花褶用于迂腐的文史、书吏或奸诈贪色的衙内、恶少。一般用绿色，绒绣散点式小碎花，纹样为花卉或八宝。武丑花褶用于武艺高强或性格诙谐的丑角人物。周身绒绣散点式纹样，多为飞禽（飞蝶或飞燕）。女花褶子用于平民家庭的女子，所谓"小家碧玉"，因其身份低，所以只用二方连续纹样作为缘饰。

1. 男褶子的文化内涵及其协调翻译

原文 Xuezi as a kind of everyday dress of ordinary people, it resemble the outer vestments worn by Buddhist monks or Taoist priests. The cloaks are rich in color, such as scholars and students normally wear blue. Black cloaks are for the poor or unsuccessful.❶

拙译 褶子作为普通人的便服，大领斜大襟，宽袖带水袖，因为像和尚、道士穿的道袍，所以俗名又叫道袍。褶子颜色丰富，一般文人和书生穿蓝褶子，下层人和不得志的人穿黑褶子。

原文 Casual Coat (Xue) is a cross-over closing coat with a cross collar，which is from the ancient wide-sleeve robe. "Xue" is a kind of casual coats for low-class people and ordinary people. Xuezi also has two styles for male and female. The male style coat has a cross collar and the coat is long to the feet. Different kinds of "Xuezi" fit different kinds of characters respectively.❷

拙译 褶子源自古代的宽袖袍服，交领对襟。褶是社会底层和普通人群日常穿用的外衣，有男女两种款式，其中男款交领，衣长至脚踝。不同款式褶子的选用要与京剧表演中不同性格的人物匹配。

❶ 赵少华. 中国京剧服饰［M］. 北京：五洲传播出版社，2004：46.
❷ 谭元杰. 中国京剧服装图谱［M］. 北京：北京工艺美术出版社，2008：93.

2. 女褶子的文化内涵及其协调翻译

女褶源自明代"小立领对襟窄袖袄",而小立领源于明代程朱理学的思想禁锢,妇女脖颈要用立领加以遮掩。京剧服装将立领袄和宽袖褶子进行改造,并融合两者特点,将内衣、外衣合二为一,创造出了京剧舞台上的小立领对襟衫——女式褶。笔者摘录了下文所示的英文表达,同时补充了拙译。

原文 Xuezi as a kind of everyday dress of ordinary people, it resemble the outer vestments worn by Buddhist monks or Taoist priests. The cloaks are rich in color. For female characters, such cloaks are actually short jackets with water sleeves attached to the cuffs. Most Xuezi are made of silk or satin, but in some isolated cases, cotton is used.❶

拙译 褶子作为普通大众的便服,大领斜大襟,宽袖带水袖,因此像和尚或道士穿的道袍,颜色丰富。对于女性角色,这样的外衣样式其实都带水袖,且尺寸较短,需下着衬裙。其通常采用绸缎制作,也有个别用布制作的褶子。

原文 Casual Coat (Xuezi) is a cross-over closing coat with a cross collar, which is from the ancient wide-sleeve robe. "Xue" is a kind of casual coats for low-class people and ordinary people. Xuezi also has two styles for male and female. the male style coat has a cross collar and the coat is long to the feet. Different kinds of "Xuezi" fit different kinds of characters respectively. The female coat is with a center front closing and it's only long to the knee , and a skirt must be dressed inside.❷

拙译 褶子属于普通的外衣,交领宽袖,为地位低下或低阶层的男女日常穿用,只是男女在款式上有所不同。男款褶子为交领,衣长至脚踝,其款式适用于不同类型的人物;女款褶子为对襟小立领,衣长仅过膝,里面一定得穿裙子。

原文 As for daily wear, the Xuezi is a simple, geometrically cut robe from Song dynasty, the Jiaolingpao (long robe). It can be worn by almost all characters in Jingju. The front and back are cut in a single, floor-length. Wide sleeves with water sleeves, and the side arms are vented to the waist. The Xuezi comes from the straight crossover closing to the right, which is finished with a wide straight band. It is the variations in the color, fabric, cut and surface embellishment that determine the different forms that the Xuezi can take, such as Huaxuezi and Suxuezi. Additional costume pieces can be added to the Xuezi for other looks. Nüxuezi is knee length and worn with a skirt. A standing band collar and a center-front closing secured with frogs. It has long straight sleeves with water attached. The center front opening Nüxuezi has different embroidery and

❶ 赵少华. 中国京剧服饰 [M]. 北京:五洲传播出版社,2004:46.

❷ 谭元杰. 中国京剧服装图谱 [M]. 北京:北京工艺美术出版社,2008:93.

borders for a variety of uses.❶

拙译 褶子作为日常穿着，在宋代通过简单的立体裁剪而成，称作交领袍（长袍），在京剧中几乎适合所有的人物穿用。前后片独立裁剪，长度及地。宽袖带水袖，侧面从胳膊一直延伸到腰部。褶子的右边直线交叉，一条宽带左压右。褶子的颜色、材质和剪裁，以及表面点缀的变化都决定了其呈现的形式，如花褶子和素褶子。在褶子上添加其他服饰时就会呈现另外的样子。女褶子长及膝盖，并有配裙，款式为立领和前开襟，并用盘扣固定，笔直的长袖和水袖相接。其刺绣图案和镶边因其用途差异，而在选择上也明显不同。

译文剖析 对于穿着褶子人物身份背景的翻译，赵将其翻译为 "a kind of everyday dress of ordinary people" 供普通人日常穿着的服装；谭将其翻译为 "a kind of casual coats for low-class people or ordinary people" 低阶层或普通人的休闲服装；亚历山德拉译为 "as for daily wear, it can be worn by almost all characters in JingJu"。在京剧中，褶子是上到皇帝下到穷人的几乎所有角色都可以穿着的服装，所以笔者认为亚历山德拉的描述最中肯，包括对于褶子的出处——宋代的斜领大袖衫，即 "geometrically cut robe from Song dynasty, the Jiaolingpao (Long robe)"。"geometrically" 在《牛津高阶英汉双解词典》中的解释为 "in a way that is like the lines, shapes,etc. used in geometry, especially because of having regular shapes or lines"❷，几何的剪裁，交领衫。并且在后面介绍褶子式样特征的时候还具体说明了 "斜领大袖衫" 的特征："the straight crossover closing to the right, which is finished with a wide straight band"，针对英汉两种语言的单词转换和语言风格的使用给予关照，方便中西文化交流。谭将其表述为 "from the ancient wide-sleeve robe"，即来自古代时期的大袖衫，相对欠详细，首先没有说明其来自古代的哪个朝代，其次 "wide-sleeves robe" 大袖衫不具有特定的专属特征，古代的服装袖子都很肥大，并不只有 "斜领大袖衫" 的袖子宽大，如此表述容易让读者产生歧义。

对褶子式样的描述，赵译表述为 "resemble the outer vestments worn by Buddhist monks or Taoist priests"，像开氅一样，再一次列举了僧袍和道袍的例子。笔者认为这样的阐述无法帮助外国读者了解其式样究竟如何。赵译文的描述为 "male style coat has a cross collar and the coat is long to the feet"，提及其交领和长及脚面，但是并没有翻译其阔袖带水袖的特征。亚历山德拉描述为 "Front and back are cut in a single, floor-length. Wide sleeves with water sleeves, and the side arms are vented to the waist"，对褶子的所有形式特征都进行了详细描述。针对其颜色和纹样，原文的内涵是褶子适用于所有阶层，但不同的颜色、面料和纹样代表了不同的身份定位。针对这一特点，赵表述为 "cloaks are rich in color, made of silk or satin, but in some isolated cases, cotton is used"；谭表述

❶ BONDS ALEXANDRA B. Beijing Opera Costumes: the Visual Communication of Character and Culture［M］. Honolulu: University of Hawaii Press, 2008: 135.
❷ 霍恩比. 牛津高阶英汉双解词典［M］. 王玉章, 赵翠莲, 邹晓玲, 等译. 7版. 北京: 商务印书馆, 2010: 850.

为 "different kinds of 'Xuezi' fit different kinds of characters respectively"；亚历山德拉表述为 "it is the variations in the color, fabric, cut, and surface embellishment that determine the different forms that the Xuezi can take, such as Huaxuezi and Suxuezi"。笔者认为，赵传递出褶子颜色和面料等方面的丰富，但没有说明这些不同分别代表了什么。谭阐释了褶子的选用需对应于不同的人物，但没有说明这些褶子在哪些地方不一样。亚历山德拉则结合了赵和谭各自表述上的优点，既说明了在纹样、颜色等使用方面的丰富和差异，也阐释了这些不同代表了不同身份的角色定位。

最后对于女褶子的表述，谭将其阐释为 "The female coat is with a center front closing and it's only long to the knee, and a skirt must be dressed inside"，亚历山德拉阐释为 "Nüxuezi has different embroidery and borders for a variety of uses"。笔者认为两位学者都对女褶子的特点给予了关注，但都欠全面，若将两位学者的表述融合，则不仅说明了女褶子在式样上和褶子一样，只是长度短至膝盖，需内着衬裙，并且传递出了其在图案的使用上有女性专属的特色。如此协调翻译更加方便外国读者非常深入地了解其文化内涵，传播中国传统京剧文化。

（九）宫装的文化内涵及其协调翻译

宫装又称宫衣，属于女用常礼服，用于皇妃、公主在一种比较随意闲适的后宫场合穿着。宫装虽然极为华丽，但规格低于女蟒，不能用于庄严隆重的环境。其形制是衣裙一体（上衣下裳相连），圆领对襟，阔袖（袖口大镶大沿，具有清代满族妇女服饰的典型特征），肥腰（革带仅为装饰），腰际以下缀有彩色飘带（3层，共计64条），内连衬裙。服色以红为主，杂以各色作为辅衬。周身满绣纹样（上衣绒绣飞凤牡丹，飘带绒绣草花或草凤）。穿戴宫装时还需要加上云肩（此种盖肩装饰品为小立领，绣花，周围饰以网子穗，也具有清代女装特征）。下面对照分析英汉两种不同语言对于宫装的描述。

原文 Under the generic name of royal apparel (Gongzhuang), a long and loose outer garment is usually worn by an empress,a queen or ladies-in-waiting on the stage. It features a ring-shaped collar, a large lapel and a loosely fitting waist, and reaches from the neck to the instep. A resplendent costume, primarily in red, it has a number of richly followed patterns on the upper part while its lower part is encircled alternately by dropping streamers of ribbon in unequal lengths. It also has s tasseled shawl or yunjian to cover it around the collar, and the sleeves are embroidered with stripes about one inch wide in silk thread, to which some two-inch-wide colorful fringes are sewn.Yang Yuhuan, the concubine of Emperor Xuanzong (reigned 712-755), wears such a garment in the scene of "Pavilion of Hundred Flowers" in *Drunken Beauty*.❶

---

❶ PAN XIAFENG. The Stagecraft of Peking Opera［M］. Beijing：New World Press，1995：118.

拙译 通用的官装属于宽松型的长款皇室服饰，京剧舞台上皇帝、皇后和宫女都穿这类外衣，并且从脖颈到脚背分别配有环形项圈，以及大翻领和宽松的腰部装饰。色泽华丽，主要选用红色。衣服上半身部分的配饰模式相对固定，而下半身部分则配有长短不一的飘带交替环绕；衣领还配有流苏、披肩或云肩；袖子上绣有约2.5厘米宽的丝线条纹，还缝着一些约5厘米宽的彩色条纹。京剧《贵妃醉酒》中唐玄宗（公元712—755）的爱妃杨玉环在百花亭独饮时就曾穿用了官装。

原文 Imperial concubines and princesses wear palace robes. Palace robes are longer and look more elegant than dragon robes for female characters, with embroidered ribbons and tassels sewn to the waist.[1]

拙译 官装是王妃和公主穿的礼服。衣裙一体，圆领对襟，肥腰阔袖，长于女蟒。周身满绣纹样，底襟有彩色绣花飘带和穗子，但不用于庄重的场合。

原文 The Gongzhuang is worn by princesses and concubines, for leisure, dancing and other activities in the inner palace. Its rank is lower than Nümang, so it would not appear for important ceremonial occasions. It is full length, and that length is composed of layers of steamers. It has rectangular shape of the bodice, round-necked. Wide sleeves are decorated with seven colorful bands with wavy edges separated by contrasting bias binding, and these bands are repeated in the streamers of the skirt era. And it is richly embroidered in a wide range of beautiful colors.The floral imagery on the streamers, aprons and sleeve bands often includes peonies, as they are emblematic of feminine virtue and beauty. The phoenix, with it's associations with imperial women, appears on the bodice and the front aprons ,embroidered with a wide range of pretty colors to depict the flamboyance of this mythological bird.[2]

拙译 官装是公主和妃嫔在内宫休闲或参加跳舞等活动时穿用的服装，其级别低于女蟒，一般不会在重要的仪式场合穿用。官装属于长款服饰，其长度与腰下端所缀的数层飘带有关。官装为圆领，腰身宽大，同样较为宽大的袖子上有七条色彩各异、波浪型的条带接续排开，这些带子的色彩同样出现在裙子的飘带上。官装的刺绣内容多样，色彩丰富。飘带、如意头大飘带和袖子上均绣有花卉，其中牡丹因其象征女性的美丽和德行而成为常见的花卉纹样。上身和正中的如意头大飘带上有象征后宫女性的凤凰图案，用色绚丽多彩以展现这种神鸟的华丽。

译文剖析 对官装文化背景的翻译，赵译解释为皇宫中妃子和公主穿着的服装，没有阐明其服装定位为只适用于休闲的场合。亚历山德拉的翻译则将官装所具有的特

---

❶ 赵少华. 中国京剧服饰［M］. 北京：五洲传播出版社，1999：24.
❷ BONDS ALEXANDRA B. Beijing Opera Costumes: the Visual Communication of Character and Culture［M］. Honolulu：University of Hawaii Press，2008：156-157.

定的文化背景和具体要求全部详细阐释"for leisure, in the inner palace, rank is lower than Nvmang, not appear for important ceremonial occasions"（用于闲暇时，在后宫服饰中等级比女蟒低，不用于正式场合）。这些对官装的文化内涵的阐释忠实于原文内容，并协调了两种语言和翻译风格。

对于官装款式的描述，圆领，腰身肥大，阔袖带水袖，亚历山德拉都给予了关注。对腰身肥大的翻译，亚历山德拉阐释为"rectangular shape of the bodice"长方形的身形，笔者认为可以借鉴阙艳华主编的《中英文对照京剧服饰术语》一书，将其协调为"fits loosely on the body"宽松的腰身，并不具体说明其形状属于哪种，避免读者产生其他联想。关于对襟这一领子上的特点，两位学者都没有做出阐释。笔者认为可以协调为"the collar closes at the front"，即领子在正前方闭合，以异化的方式解释对襟的概念。另外，官装的样式比较复杂，特别是水袖部分有彩绣，并且身上有彩色飘带，赵的译文虽忽略了袖子的特色，但将彩色飘带处的细节表述为"embroiedered ribbons and tassels to the waist"，精练恰当，也不会让读者产生误解。亚历山德拉的描述则相对更为详尽，先对袖子上的彩绣协调翻译为"sleeves are decorated with seven colorful bands with wavy edges separated by contrasting bias binding"，然后又补充了裙子上的彩色飘带是袖子上那些条带的再一次使用（and these bands are repeated in the streamers of the skirt era）。笔者认为关于袖子的描述很恰当，但裙子上的彩色飘带和袖子的彩绣并不是同一种东西，袖子上的彩绣相对固定，而裙子上的飘带有三层、带穗且飘动。这样的表述似乎巧妙，但并没完全忠实于官装的文化。所以笔者认为对袖子上彩绣的描述可以借鉴亚历山德拉的翻译，但对裙子上彩色飘带的描述则可以借鉴赵译文中的相关描述，补加对飘带层次和总量的表述："embroiedered three layers ribbons as many as sixty four in total, and tassels to the waist"。

穿着官装时，需要内着衬裙，外穿云肩，并且官装的颜色基本为红色，很少有其他颜色。这些特点两位学者并没有关注，为方便外国读者了解其文化内涵，笔者认为可将其补充协调为："It is generally red, and when wear Gong zhuang, need wear a pleated skirt inside, and a cloud-shaped cape attached to the shoulder."

（十）古装的文化内涵及其协调翻译

梅兰芳早年编演《嫦娥奔月》《天女散花》等新剧时，参考古代绘画、雕塑中的人物造型特点，大胆创制了许多新颖而别致的戏装，当时将它们统称为古装，以区别于一般通用的戏装。如《洛神》中的示梦衣、戏波衣，《嫦娥奔月》中的采花衣，《太真外传》中的舞盘衣、骊宫衣，《木兰从军》中的木兰甲，以及《天女散花》中的云台衣等，皆为梅先生所创。

以云台衣为代表的古装，本是为了在戏曲舞台上发挥古代歌舞特点而创制，显示了前辈艺术家敢于突破传统的革新精神。从形制上看，它改变了传统的服装造型：衣裳剪裁合体，上衣短如褶，裙子系在衣外，以带束腰，袖子也由肥阔而变为窄瘦。这种服式无疑使得剧中人物具有修长、窈窕的美感，符合现代审美情趣。此云台衣包括以下几个部分：上衣为淡青色古装袄子，外披直大领云肩，下裳为腰裙，胯部系小腰裙（即小侉子），腰系黄丝绦，挂玉珮，肩部垂两条风带。不妨从下列英汉对照版中再次领略古装之韵味。

原文 Unlike most of the conventional costumes in Peking Opera, the antique dress or Guzhuang, again introduced by Mei Lanfang, has so far remained a simple but elegant article of attire for actresses who play mythical or fictitious characters like the Heavenly Maid (Tiannü) in *Heavenly Maid Scatters Flowers*, the Moon Goddess (Chang'e) in *Lady chang'e Flees to the Moon* and Lin Daiyu in *Daiyu Buries Flowers*.[1]

拙译 梅兰芳再次推出的古装，不同于京剧中其他的传统服饰，女性古装简洁优雅，常被扮演神女或天女的演员穿用，如《天女散花》《嫦娥奔月》以及《黛玉葬花》之类的人物角色。

原文 The Guzhuang ("ancient-style" dress) has many manifestations, the most typical version of which is generally composed of several pieces, with an inner layer of jacket and skirt and an outer layer of variously shaped collar and peplum pieces. Most sets of garments with short peplum pieces worn over the skirt fall into this category of costume. The jacket is worn tucked into the skirt and has a lapped closing with a round neck or a center-front closing similar to that of the Nüpi. The sleeves are tubular and may or may not include water sleeves at the hem. The skirt is unusual in that it is sewn shut rather than being made of two separate panels like the standard skirt. The pleats can be formed in more than one way, either with knife pleats all going in the same direction around the skirt, or a central panel in the front with knife pleats folded towards the back on either side.

The jacket and skirt of the Guzhuang can be made of silk chiffon that is lined with a regular-weight silk fabric in a matching color. The two layers of fabric are intended to complement the flirtatious movements of the huadan roles or dancing characters. Matching accessories are worn over the jacket and skirt and include a cloud collar or bib with tabs to the waist, a belt, and a peplum with an apron and streamers.[2]

拙译 古装（古式服饰）形式各异，其特色是由几件衣服组成，内层为上衣和长

[1] PAN XIAFENG. The Stagecraft of Peking Opera［M］. Beijing：New World Press，1995：119.
[2] BONDS ALEXANDRA B. Beijing Opera Costumes：the Visual Communication of Character and Culture［M］. Honolulu：University of Hawaii Press，2008：189–190.

裙，外层为形式各异的云肩和装饰性的裙片。大部分古装都有短裙相配，上衣塞在裙子里，领子为圆领或与女帔类似的对襟长领。袖子为窄袖时，袖口不缀水袖，袖子为大袖时则接水袖。古装的长裙较为特别，它是缝合而成，并非像普通裙子那样由两片独立衣片制成。长裙的褶裥也有多种制作方式，有的褶裥以相同方向绕群一周；有的则是裙子正中间为一整片光面，称为马面，褶裥分别朝两侧折叠。

古装的上衣和长裙通常选用绉缎、薄纱制作，衬里使用常规的丝绸且颜色相配织物。双层布料的使用旨在方便花旦或舞蹈者动作的展示。上衣和长裙之外穿戴包括云肩与带有扣襻的腰箍、腰带、飘带在内的配件。

（十一）裙、裤、袄的文化内涵及其协调翻译

京剧行头在裙、裤和袄的总称中又划分出、裙袄、裤袄、竹布裤褂、罪衣裤和彩旦褂子等。其样式为上衣下裙或上衣下裤，上衣（袄子或褂子）为小立领、大襟（其中罪衣为小立领、对襟），衣长至臀部。裙、裤、袄周身纹样以花卉散点或枝子花为主。裙长从腰部下至足面，形有筒裙和百褶裙。裤为中式肥腰缅裆裤，裤长从腰部下至足面，为散腿。其中彩旦褂子有立领或圆领两种，周身宽松，衣长至臀部。

裙可以分为四种：①腰裙，一般为旦角用的衬裙，以白色居多，老旦的衬裙为绿色。白色腰裙系在衣服外面，称为"打腰包"，用以表现病人、行路人或犯人。②水裙，即白布短裙，多与茶衣、短跳结合使用，为渔民、樵夫、店伙等角色所用。③战裙，类似靠的下甲，与袄裤结合使用。④花旦用的裙子，常与袄合穿，花式很多，过去也属于入时打扮。裙袄——作为民间少妇或少女的服饰穿着时，一般要配上饭单、四喜带。

女短袄为立领，大襟，袖口不带水袖，用绸缎制作，花样和色彩多种多样。裤袄常为小户人家的年轻妇女或大户人家的丫鬟穿用。小户人家的姑娘穿裤袄时，腰里大都系着四喜带或一条汗巾。武旦用的裤袄，立领，对襟，束袖，缎制绣花，多用于民间会武艺的女子。花旦用的裤袄，清末以来才兴起，属于入时打扮，变化很大。现在一般都穿立领，大襟。小袖的绣花裤袄常加坎肩或饭单，系绣花汗巾、四喜带，为小户人家女子及丫鬟使女所用。彩旦用的裤袄，多肥大，长及膝，深色，镶阔边，多为扮茶婆、媒婆时所用。

罪衣罪裤为立领，对襟，红布制，一般为罪犯专用。

裤子的花色与上衣同。丫鬟穿着裤袄时，需配上坎肩，腰系腰巾子。而京剧的内衬服装——彩裤，各行角色均用，有红、黑、白、粉红、湖蓝等颜色，多为素地绸制。

原文 Jackets (Ao) worn by female characters have straight collars and a right-buttoning front. The cuffs are loose and no water sleeves. Such jackets are all made of satin or silk. Diverse

94

in style and color. They are worn by young women in families of limited means or maids in big, powerful families. While wearing a jacket of this kind of limited means often has an embroidered belt tied round her waist——sometimes just a long, embroidered handkerchief or towel.[1]

拙译 女式短袄立领右衽，袖口较肥且无水袖，选用绸缎制作，款式和颜色多样。袄常为小家碧玉或大户人家的小丫鬟穿用，穿时腰系四喜带，有时候用绣花腰巾或汗巾替代。

（十二）旗袍的文化内涵及其协调翻译

旗袍形成于20世纪20年代，有部分学者认为其源头可以追溯到先秦两汉时代的深衣，是悠久的中国服饰文化中最绚烂的现象和形式之一，也是中国女性的传统服装，被誉为中国国粹和女性国服。在京剧舞台上，旗袍样式更加接近清朝旗人妇女的服装样式，但尺寸略长一些，正如下列关于旗袍的英汉对照描述。

原文 "Qi" is an ethnic tag the Han Chinese use for Manchurians, who ruled China during the Qing dynasty (1644−1911). "Pao" means robe. The majority Han Chinese, sometimes called "Mandarian", slowly took on and adapted the Qipao(roughly pronounced "chee pow") into their own traditional mandarin gown. In China today, the Qipao is a close-fitting women's dress with a high neck and slit skirt. The Qipao is a close-fitting women's dress with a high neck and slit skirt. The Qipao used in Peking Opera performances are very much the same as the original Manchurian gown in its shape. The major difference is that it is a bit longer.[2]

拙译 旗是汉族对清代（1644—1911）统治中国的满族人使用的族裔标签。袍意即长袍。大多数汉族人，有时称为"说官话的人"，慢慢地将旗袍（大致发音为旗袍）改成他们自己的传统服饰。时至今日，旗袍成了一款高领开衩、合身的女式连衣裙。京剧中穿用的旗袍与原来满洲服在形制上非常相似，主要区别在于旗袍稍长一点。

原文 The Qipao used for traditional Jingju costumes is a descendant of the Qipao worn as everyday dress during the Qing dynasty by the Manchu women. The Qipao represents a significant difference in silhouette from most garments worn by Han Chinese women onstage and off. Han women's clothing consisted of two parts, a short robe or jacket and a pleated skirt, while the Qipao was a single full-length garment. The Qipao is a narrow trapezoidal-shaped dress with tubular sleeves. It closes to the right with an S-shaped curve and frogs typical of Manchu styling. A standing band collar finishes the neckline. The side seams are open to the top of the thigh. A

---

❶ 赵少华. 中国京剧服饰［M］. 北京：五洲传播出版社，2004：78.
❷ 同❶62.

slimmer, more shapely form of the Qipao continues to be worn today by Chinese women who have a taste for classical styles.

The Qipao entered the traditional Jingju costume vocabulary towards the end of the Qing dynasty, along with other Manchu-styled clothing used to identify women from ethnic minorities. The contrasting silhouette provides a readily identifiable clue of the origins of the character onstage. The onstage convention of wear reflects that of real life to some extent, with Han women in traditional Jingju retaining the two-piece garments of Han women in history and women of all other groups dressed onstage in the one-piece garments of the Manchu women of the Qing dynasty.

The Qipao is worn for everyday scenes onstage by ethnic minority women of higher status and their maids. The Qipao comes in many colors, and the designs on the surface distinguish the roles. For higher-status women, the Qipao is embroidered with sprays of flowers and sometimes butterflies or birds. In addition, the sleeve cuffs, neckline, lap, side vent and hem edges may have contrasting color borders.

The Qipao, like other forms of Manchu dress onstage, is worn with the Huapendi, footwear developed in the Qing dynasty to imitate the bound feet of the Han women. Hand movements in this robe are distinct, as the Qipao do not have water sleeves, so these characters carry a handkerchief instead. Straight legged trousers with bands at the cuffs are worn under the costume.[1]

拙译 用于传统京剧服饰的旗袍是清朝满族妇女日常服饰的变异。多数汉族妇女在舞台上穿的旗袍和生活中的旗袍在轮廓上存在明显差异。汉族女性的服装由两部分组成、短袍或衫、袄衫和百褶裙，而旗袍则是件独立的长裙，有着瘦窄的梯形板型和窄袖。从右侧观察，廓型呈现S形，并配有满族风格的盘扣。脖颈上是立领，从大腿上面位置的两侧开衩。时至今日，旗袍依然为有着传统品位的窈窕淑女穿用。

到了清朝末期，旗袍和其他满族风格的服饰一起，作为区分少数民族妇女身份的标志进入传统的京剧服饰词汇，其对比鲜明的轮廓成为舞台上了解人物角色出身的线索。舞台上的习惯穿用在一定程度上反映了现实生活中的情况，在传统京剧中，汉族妇女在舞台上依然保留着历史上"上衣下裳"的两件模式，而其他各族都遵循清朝满族妇女"一件式"旗袍的传统。

在京剧中，旗袍适合于身份较高的少数族裔女性和她们的女佣在日常生活中穿用。旗袍颜色丰富，表面的图案设计成为区分人物角色的标志。对于地位较高的女性来说，旗袍上常绣有花朵，有时还绣有蝴蝶或鸟。此外，袖口、领口、侧面和下摆边缘都有不同颜色的滚边。

旗袍与京剧舞台上其他形式的满族礼服一样，通常搭配花盆底鞋。花盆底鞋始于清代，是对汉族妇女缠足的模仿。因为没有水袖，穿上旗袍后，演员手部的动作就格

❶ BONDS ALEXANDRA B. Beijing Opera Costumes: the Visual Communication of Character and Culture［M］. Honolulu: University of Hawaii Press，2008：181.

外明显，所以剧中人物都拿着手绢。旗袍下都穿着打绑腿的直筒裤。

（十三）富贵衣的文化内涵及其协调翻译

富贵衣是京剧戏曲舞台上扮演原本穷困潦倒，但日后金榜题名、显达富贵之人所穿用的服装，故又称"穷衣"。同时，富贵衣作为传统戏曲服装中吉祥的预兆，一般放在衣箱的最上面，在传递吉祥的同时还可以保护箱子下面用金或银线绣制的珍贵服装。例如，《玉堂春》中的王景隆，落魄后被赶出妓院时，穿的就是富贵衣;《金玉奴》中的莫稽，先穿富贵衣，后来得中状元，成为钦差八府巡按;《红鬃烈马》之《彩楼配》一折中，薛平贵也是先穿富贵衣，后来才得唐室天下，做了皇帝。下面的英汉表达从不同角度阐释了其内涵。

原文 The Fuguiyi is a casual costume in Peking Opera. It is basically a Xuezi robe that overlaps at the front with many small silk patches of various colors to show that it is a heavily patched rag. This kind of costume is usually worn by poverty-stricken, depressed characters. It is known as a robe of fortune, because at length fortune will deliver these characters from their bad situation.[1]

拙译 在传统京剧中，富贵衣属于便装。一般是前部交叉重叠的褶子袍服，打着许多不同颜色的丝绸小补丁，给人以补丁摞补丁和破衣烂衫的感觉。富贵衣为遭受贫困打击、情绪沮丧的人物穿用，之所以被称为富贵衣，是因为这些人物最终都会摆脱困境。

原文 The Fuguiyi is ironically called the garment of wealth and nobility. Although it is covered in patches and worn by those in impoverished circumstances, it nevertheless projects the audience's wish that the character will come to better times. It has rectangular sleeves, and crossover diagonal closing, with a band at the neck that can be white or black with white piping around the edge. Fuguiyi is made of silk and carefully embellished with colorful silk patches, but in real life, patches are used to cover satins or rend. The male character wearing the Fuguiyi wears a soft, black hat.[2]

拙译 富贵衣有种说法是对具有财富和权贵意义服装的嘲讽。尽管富贵衣上缀满补丁，为处于贫困的人群穿用，但是它能够表达观众对角色人物美好未来的祝愿。富贵衣有着长方形袖子，交领，领口有白色或黑色的细长飘带。京剧戏曲中的富贵衣由

---

[1] YI BIAN. Peking Opera: the Cream of Chinese Culture [M]. Beijing: Foreign Languages Press, 2005: 54.

[2] BONDS ALEXANDRA B. Beijing Opera Costumes: the Visual Communication of Character and Culture [M]. Honolulu: University of Hawaii Press, 2008: 116–117, 略有改动.

丝绸制成，并精心点缀着多彩的丝绸补丁。但在现实生活中，补丁要打在缎子或撕裂的口子上面。男性角色穿富贵衣时，都配有一顶材质柔软的黑帽。

原文 In actual fact, Fuguiyi is just the patched garment, which is nothing but a black commoner's coat with a number of variegated patches stitched on. In this case, not only does it impart a sense of shabbiness, but it also implies a sort of the wearer, in powerful as well as in wealth, hence the name in Chinese name Fuguiyi. It conveys, as it were, a broad hint that poverty is transient whereas social mobility is ever-present.❶

拙译 实际上，富贵衣是补丁衣，只不过是普通人穿的黑布衣，上面补丁摞补丁。在传统京剧中，富贵衣在传递衣者当时窘境的同时，也暗含着其潜在的财富和力量象征，因此称其为富贵衣。富贵衣暗指贫穷是短暂的，而社会变化则是永恒的。

第三节
京剧大衣箱饰物配件文化内涵及其协调翻译

京剧大衣箱中的饰物配件主要有坎肩、斗篷、蓑衣、领衣、饭单、四喜带、偏带、玉带、丝绦、腰巾、腰箍、云肩、水袖和水裙等，分别简述如下。

（一）坎肩文化内涵及其协调翻译

原文 大坎肩又称"马甲""背褡""背心"，是加饰在袍外的无袖上衣。有花、素两种，分男、女两式。男式大坎肩为直领对襟，由黑色或古铜色缎料制面，四周用平金绣绣篆文、古钱或万字图案，腰系丝绦。主要为身份较低的学究、私塾先生、江湖医生穿用。女式大坎肩为直领对襟，衣长过膝，两侧开衩。通常由缎料制面、金丝盘边、彩绒绣角，服色有黑、白、粉、湖蓝、果绿等，主要为宫女、侍女穿用。还有一种对襟、衣长过膝的黑缎女坎肩，专用于老旦角色，穿在女老斗衣之外，领口镶白色云头或绣圆型"寿"字，腰系素腰巾或丝绦。❷

译文 Dakanjian (Long Sleeveless Jacket) is an outer garment, also called Majia,

---

❶ PAN XIAFENG. The Stagecraft of Peking Opera [M]. Beijing: New World Press, 1995: 116–117, 略有改动.

❷ 阙艳华, 董新颖, 王娜, 等. 中英文对照京剧服饰术语 [M]. 覃爱东, 陶西雷, 校译. 北京: 学苑出版社, 2017: 40, 略有改动.

Beida and Beixin. There are different types for male and femal characters, either in a richly-decorated style or in a plain style. The male type has a straight collar with a front opening, made of black or bronze silk with Pingjin embroidery (one of the traditional Chinese embroidery crafts) in the shape of seal script, an ancient coin or the swastika, and the costume also comes with a silk ribbon belt. It is worn by pedantic scholars, private tutors or quack doctors. The female type also has a straight collar with a front-opening, and is knee-length. It can be made of black, white, pink, turquoise blue, fruity green: all various colours. The costume is embroidered with grasses and edged in a variety of colours. It is a type of long clothing for female servants, to be worn over the Xuezi. Another type of sleeveless jacket or kanjian is exclusively used by elder female characters, with white cloud shape or the Chinese character "shou", meaning longevity in a round shape around the neck. Worn over a kind of gown for elder women called Nülaodouyi, It is tied with a silk ribbon or a waistband of a plain colour.[1]

原文 小坎肩分男、女两式。男式小坎肩为立领大襟，上缀一排铜盘扣，主要为寄人篱下的角色穿用。女式小坎肩为立领大襟，衣长及臀，彩色缎料制，周身彩绣花卉等图案，加穿于短袄外，主要为宦门富室人家的侍女穿用。[2]

译文 Xiaokanjian (Short Sleeveless Jacket) is a sleeveless outer garment worn by those who are dependent on other people for a living. It is made of black satin material with a stand-up collar and a side opening, with a row of copper buttons. There is also a female style, which is made of coloured satin with stand-up collar and side opening, and is hip-length. It is worn by maid servants over their short padded coat.[2]

原文 道姑坎肩为对襟直领，左右开衩，衣长过膝，通身素洁，基本无任何纹饰。主要为女道姑穿用，穿时不衬内裙。[3]

译文 Daogu Kanjian (Sleeveless Coat for Nuns), as the term suggests, is worn by Taoist nuns. It has a front parallel opening, a straight collar, side slits, It is knee-length and plain.[3]

（二）斗篷文化内涵及其协调翻译

斗篷作为挡风御寒的服装，在戏曲舞台上常被穿用，按照其使用功能可分为三类，

---

[1] 阚艳华，董新颖，王娜，等. 中英文对照京剧服饰术语［M］. 覃爱东，陶西雷，校译. 北京：学苑出版社，2017：40，略有改动.

[2] 同[1]42，略有改动.

[3] 同[1]46，略有改动.

京剧服饰大衣箱文化阐释及翻译协调美

第三章

99

即长斗篷、短斗篷和防雨用的蓑衣；按照其款式又可分为男、女和花、素。男、女长斗篷均配有风帽。在纹样上，女斗篷常用花卉鱼鸟，而男斗篷通常分为红素斗篷和绣龙斗篷。短斗篷主要体现在新编历史剧中，蓑衣在样式上同短斗篷，随周身横排数道垂穗，类似南方渔夫的防雨用具。

原文 长式斗篷，又称"一口钟"，是披在肩上的无袖外衣，衣长约1.3米，下宽大如钟。小立领，缝襻带，多与同色风帽配套穿用。男斗篷的服色有红、黄两种；女斗篷的服色较多。常用在挡风御寒及出门上路时。❶

译文 Changshidoupeng (Long Cloak) is a sleeveless outer garment draped over the shoulders. It is about 133cm in length with a loose fting, bell-like lower hem. It has a small Mandarin collar, a hooped neckline, and usually has a hood of the same colour. Male characters wear red and yellow, whilst the female characters can use a much wider vanity of colours. On stage, long cloaks are used to show that the character needs to protect themselves from tie cold wind. The cloaks for female characters are embroidered with various designs, such as a peacock spreading its plumage, and flowers of all seson. ❶

译文剖析 此处的译文运用了增译法，充分考虑译文读者对原文的理解与接受能力，协调文化差异，增补必要的文化信息，以便更好地理解女斗篷的特征。

原文 短斗篷样式与长式斗篷相同，衣长过膝，绸料制，素面无绣，前衣襟与底摆均镶白色毛皮边，主要为剧中番邦异族所穿。❷

译文 Duanshidoupeng (Short Cloak) is similar to the long cloak in style, is hip length, and features no embroidery. It also has a white fur brim around the front and bottom sections. It is worn by people from foreign lands.❸

（三）蓑衣文化内涵及其协调翻译

原文 蓑衣形似短斗篷。圆领，领口有襻带，衣长及臀。衣上镶饰横排丝线穗，一排压一排，层次匀密，具有飘动自如的特点。服色有秋香、咖啡、皎月等。❷

译文 Suoyi (Alpine-rush Rain Cape) is made of alpine rush and used in southern China to protect farmers from rain. It is brown in colour, and looks like a cloak with a round collar and belt around the collar. It is hip-length.❷

❶ 阙艳华，董新颖，王娜，等. 中英文对照京剧服饰术语［M］. 覃爱东，陶西雷，校译. 北京：学苑出版社，2017：42，略有改动.

❷ 同❶44，略有改动.

❸ 同❶44.

译文剖析 京剧艺术表演中的蓑衣属于类似短斗篷一样披在肩上的无袖外衣，其材质并非是真正的蓑草，而是利用数层横排的丝线，以带给观众生活中防雨用具蓑衣的视觉效果。译文中对蓑衣的解释"made of alpine rush and used in southern China to protect farmers from rain"，在一定程度上偏向于生活中真实的蓑衣。可以尝试将其补充为"made of somethings looked as if it were alpine rush"，既保证了其层次匀密的特点，也将丝线穗飘动自如的效果蕴含其中，以方便读者理解。此外，在对剧装蓑衣色彩的翻译中，译文仅以"brown"概括"秋香、咖啡、皎月"等色，似乎不足，笔者拟将其补充协调为"chartreuse, brown, bright moon colour"，完善对蓑衣原貌的呈现。

（四）领衣文化内涵及其协调翻译

原文 领衣为天蓝色缎料制，呈圈形硬领状。领口下连领身，领身为舌形、对襟、齐腰的缎片，是戏曲服装中象征少数民族的装饰物。❶

译文 Lingyi (Decorative Collar) is a decorative accessory worn around the neck. It is made of blue satin. The bottom of the decorative collar is connected to the body of the collar on the costume by means of a piece of satin in the shape of a tongue. The decorative collar covers the chest, the end of which is tucked into the belt.❶

（五）饭单文化内涵及其协调翻译

饭单（围裙、肚兜）源于中南地区农家妇女生活，是民间少女的花兜肚，一般为黑色，用于胸部和腹部，兼有围裙的用途。饭单又分为大饭单、小饭单两种。大饭单长及膝，专用于穿青褶子的贫家妇女；小饭单长及胯，专用于穿裤袄的民间少女。在传统京剧表演中，这类配饰为穿袄裤和配坎肩的花旦（小家碧玉、丫鬟）专用。中外学者对此也有类似描述，笔者同时将拙译呈现如下。

原文 The apron is used to indicate working women. The large apron (Dafandan) is worn over a Nüpi or Nüxuezi by the Qingyi roles in lower circumstances or when at work, and the small apron (Xiaofandan) is worn by the Huadan, often with the jacket and trousers.❷

拙译 饭单用来指代劳动妇女。大饭单为身处较低地位或劳作的青衣在其女帔或女褶子外穿用；小饭单是花旦搭配上衣和裤子时穿用。

❶ 阙艳华，董新颖，王娜，等. 中英文对照京剧服饰术语［M］. 覃爱东，陶西雷，校译. 北京：学苑出版社，2017：45，略有改动.

❷ BONDS ALEXANDRA B. Beijing Opera Costumes: the Visual Communication of Character and Culture［M］. Honolulu: University of Hawaii Press, 2008：195.

原文 上端尖圆，顶部有扣襻，与上衣领连结。由多种颜色的缎料或丝绒制成，上绣枝子花图案。饭单有大饭单和小饭单之分。大饭单专用于穿女青褶子的妇女，兼有围裙的用途；小饭单主要为贫民少女穿用。❶

译文 Fandan (Apron-like Outerwear) is an accessory similar to the so-called belly-band, made of satin of various colours or velvet, and wom by little girls in common families. It also has the function of an apron, With a buckle, a connective to attach the fandan to the collar, and buttons on the top. It should be longer than the short Chinese-style jacket called the "Ao".❷

（六）四喜带文化内涵及其协调翻译

四喜带俗称马面，古称蔽膝，是绣花腰带，下端镶穗，束于袄外。传统戏曲舞台上扮演小家碧玉、丫鬟等都可以穿用。色彩不限，一般上部为如意头形装饰纹，下部彩绣枝子花，但不宜绣凤图案。其汉英对照表达如下。

原文 "四喜带"俗称"马面"。正面绣花，下端镶穗，束于袄外。主要为穿袄裤和配坎肩、饭单（围裙、肚兜）的花旦（小家碧玉、丫鬟）穿用。❸

译文 Sixidai (Fringed Belt), literally "four happiness belt", is a decorative accessory also named Mamian (Horse Face). It is an embroidered belt with floral designs and a fringed hem, and is tied outside the coat. Specially designed for vivacious young female roles or Huadan, it is usually worn with a sleeveless jacket, lined trousers and a Fandan (aprons and undergarment covering the abdomen).❹

（七）偏带文化内涵及其协调翻译

原文 因古装衣的上衣和大裙颜色单纯，故在腰部右侧佩饰一条垂直的彩绣飘带，叫作"偏带"。偏带与色彩艳丽的小裙（二裙）、云肩、腰箍等一起组成颜色、花形各异的、能衬托出古代妇女丰姿的古装衣。❺

译文 Piandai (Embroidered Colourful Belt) is an accessory to add colour to more

---

❶ 阙艳华，董新颖，王娜，等. 中英文对照京剧服饰术语［M］. 覃爱东，陶西雷，校译. 北京：学苑出版社，2017：54.

❷ 同❶54，略有改动.

❸ 同❶53.

❹ 同❶53，略有改动.

❺ 同❶48.

simple and plain looking jackets and long skirts. It's a newly made"ancient costumes" (Guzhuangyi) by Mei Lanfang. It is an embroidered belt with colourful patterns worn on the right side of the waist. Together with a brightly coloured Xiaoqun (small skirt), a decorative cape called Yunjian (cloud-like shoulder) and a waist belt to form the Guzhuangyi, which can describe the femininity of ancient Chinese women.❶

译文剖析 此处的译文巧用增译法，协调译文与原文的差异，补充了相关信息，如梅兰芳先生早年编演《嫦娥奔月》《天女散花》等新剧时，参考古代绘画、雕塑中的人物造型特点，大胆创制了许多新颖而别致的戏装，被称为"古装"，形成了一种适应面大的通用服装，并成为定式而保留下来。

（八）玉带文化内涵及其协调翻译

原文 玉带以竹片制骨，外蒙色缎或黑绒，带面上镶有称为"品片"的方形或圆形硬片。品片上嵌有雕龙、篆书等图形。舞台上的玉带和生活中的腰带相近，但不同的是生活中的腰带紧箍在身上，而舞台上的玉带则松松地挂在腰部，起到配合表演和装饰的作用。玉带使用时常与蟒、官衣搭配。❷

译文 Yudai (Jade BelIt) is an accessory that is tied around the waist when characters wear imperial robes (Mang) for official robes. It is covered by black satin or velvet cloth and the surface of the belt is embroidered with square or round tablets named Pinpian, which has various kinds of jade decorations with different patterns. It does not serve real function and is worn only for decoration.❸

（九）丝绦文化内涵及其协调翻译

丝绦是丝编的带子或绳子，缠于腰间。通常搭配道袍、围裙等衣物。在中英文中都有类似阐释。

原文 The silk sash, as a waist ornament for performer, is especially useful for energetic displays. Actors, more often than actresses, appear to start fighting (Qida) or to walk sideways around the stage (Zoubian) with kicks or hurls of a sash (Tiluandai or Tidadai❹), so as to make

❶ 阚艳华，董新颖，王娜，等. 中英文对照京剧服饰术语［M］. 覃爱东，陶西雷，校译. 北京：学苑出版社，2017：48，略有改动.

❷ 同❶60.

❸ 同❶60，略有改动.

❹ 原文为"Shuailuandai"，经请教北京京剧院主任舞台技师、剧装专家彭仲林先生后修改为"Tidadai"。

their rapid gait all the more energetic and striking.❶

拙译 丝绦作为腰饰，有助于体现演员的活力四射。多数情况下是男演员，在开始打斗之前（起打）或沿着舞台走圈儿（走边）时踢打着绦带（踢鸾带或踢大带），这样可以使他们更加步态轻盈，表演更具感染力。

补译 走边是京剧表演程式，展示了身怀武艺的剧中人轻装潜行，因怕人看见而多在墙边、道边潜身夜行，常用于侦察、巡查、夜行、暗袭或赶路等特定情境。其表演动作灵敏，节奏明快，在轻巧中见稳重。走边的动作主要由正、反云手，各种踢腿和飞脚、旋子、蹦子、扫堂腿、飞天十响以及快三步等组合而成；并且分为单人、双人和多人等多种表演形式。演员手持或身带兵器的走边相对比较复杂，常伴以全堂锣鼓。如果间以曲牌，演员连唱带舞的称作响边；只用轻轻的堂鼓声的称为鼓边、哑边；表现水上行舟，用小锣突出水声效果的又被称作水边。

拙译 As a performing mode of Beijing Opera, Zoubian shows how martial masters of agility creep sideways along the wall and the street for fear of being caught. Zoubian is widely applied in certain circumstances like detection, patrol, sneak attack and tasks executed at night. Its movements feature its agility, fast rhythm and a combination of agility and decorum. Elementary movements of it are composed of waving the hands like clouds, kicking a leg, flying kicks, somersaults, Bengzi, circular foot sweep, Feitianshixiang (actors clapping certain parts of their own body rapidly and continuously followed by ten loud clapping sound). Zoubian can be performed by either one single person or a group of actors (no less than two people). It will be rather complicated if actors have weapons in hand or with them. In this situation, actors walk sideways around the stage accompanied by loud beats of drums and gongs. If Qupai tunes are threaded into the beating of the drums and gongs, while actors inject dancing and singing into the movements, then Zoubian of this kind can be called Xiangbian. Zoubian with light drum beats is named Gubian or Yabian. Zoubian is specifically called Shuibian when gongs are used to imitate the sound of water to image a scene on board.

原文 "丝绦"俗称"绦子"，用丝线或棉线编织而成，全长约两丈，取中系八宝襻，结下有一条两尺半长的练穗。使用时，先依男左女右的原则，将穗定位在后腰部，再将左右绦绳顺时针绕至前腰正中部位，系蝴蝶结固定。主要为文人、谋士、师爷、塾师、秀才使用，亦作皇后、王妃的素服腰饰。❷

译文 Sitao (Silk Belt) is an accessory commonly known as Taozi (ribbon). It is a rope knitted from silk or cotton, 6.67 meter in length, which has two silk fringes 83 meter in length, which has, at each end. It is fixed around the back of the waist according to Chinese

❶ PAN XIAFENG. The Stagecraft of Peking Opera [M]. Beijing: New World Press, 1995: 110.
❷ 阚艳华，董新颖，王娜，等. 中英文对照京剧服饰术语 [M]. 覃爱东，陶西雷，校译. 北京: 学苑出版社，2017: 47.

custom of "male wears it on the left, female on the right". Then the two fringes are tied around the front of the waist in a bow. Scholars, counsellors, private advisers and private teachers have a silk belt, and queens and princesses in plain clothes also have a silk belelt for decoration.❶

（十）腰巾文化内涵及其协调翻译

腰巾是穿裤、袄的花旦演员使用的装饰品。色调、绣花、式样与生活中的大汗巾儿相似。用它束腰后，还要留半截耷拉下来，这样会显得更加俏丽。下面的中英文对照阐释将从不同侧面阐释其文化内涵。

原文 "腰巾"是由单幅绸料制成，长约八尺，有绣花腰巾和素色腰巾两种。绣花腰巾又称"绣头腰巾"，有红、粉、湖蓝、皎月等颜色，两端彩绣草花及水纹图案。素色腰巾有蓝、白等色。蓝色的束于青衣腰间；白色的加饰于衣外，表示挂孝。❷

译文 Aojin (Waist Band), a decorative accessory made of silk, is 2.67 meters in length. There are two types: plain and embroidered ones. The embroidereones are red, pink, lake-blue and bright moon in colour, and feature designs of grass, flowers anwater on both sides. The plain bands have much less colour: blue and white. The blue one is for the young lead actress and the white is worn for characters in mourning.❸

（十一）腰箍文化内涵及其协调翻译

原文 这是一种以光缎为面料的绣花宽片腰带，扎在穿古装角色的人物腰部。使用时，颜色、图案必须与人物穿用的服装相配套。❹

译文 Yaogu (Waist Girdle) is a decorative accessory. It is a wide, embroidered waist belt made of glossy satin. Both the colour and the foral designs on the belt should match the garment the performer wears on the stage.❺

（十二）盖头文化内涵及其协调翻译

原文 盖头的样式呈正方形，四角镶穗。由红色缎料制成，彩绣凤形或花型图案。为

---

❶ 阙艳华，董新颖，王娜，等. 中英文对照京剧服饰术语［M］. 覃爱东，陶西雷，校译. 北京：学苑出版社，2017：47.
❷ 同❶52.
❸ 同❶52，略有改动.
❹ 同❶51.
❺ 同❶51，略有改动.

105

新婚女子遮盖头部的装饰物。❶

译文 Gaitou (Head Cover), literally "cover the head", is a square kerchief made of red satin with fringes on four ends, used for the bride-to-be to cover her face and head. A design of a phoenix is embroidered on it.❷

（十三）云肩文化内涵及其协调翻译

原文 "云肩"是一种隋朝以后发展而成的女性衣饰，样式为立领、对襟，周围饰有穗子。到清代时，云肩已普及社会的各个阶层，多在岁时节令或婚嫁时佩戴。戏曲舞台上，演员佩戴的云肩十分华丽，有繁复的绣花图案。❸

译文 Yunjian (Decorative Cape), literally "cloud-like shoulders", is a decorative piece of female clothing. It is worn around the neck and supported by the shoulders. On stage it is an accessory with elaborate and splendid embroidery and tassels along the lower edge.❹

（十四）水袖文化内涵及其协调翻译

原文 水袖是缝在蟒、开氅、帔、褶子以及个别短衣（如茶衣）袖口上的一段长方形白色纺绸。因其甩动时形似水波，故名"水袖"。水袖的长短随演员身材的高矮、行当的不同及剧情的需要而定。男用水袖长约一尺许（丑角用较短），女用水袖长约二尺。水袖的来源众说不一，以参照明代服装衬袍出袖（套袖）的形式衍化而来的说法较为普遍。水袖的功能随着京剧艺术的发展由简到繁，并已成为一种纯粹的表演和舞蹈工具。❺

译文 Shuixiu (Water Sleeves) is a piece of rectangular white silk sewn on the sleeves of dragon robes, everyday dress, warrior cloaks, ordinary cloaks and individual Jackets. They almost look like the billowing waves of water when manipulated, hence the name-water sleeve. The length of the water sleeves depends on the actor's height, role and play. For instance, the water sleeves for female roles are one foot long, for male roles are two foot long, up to a maximum of three feet. Absorbed into Peking Opera in the late twentieth century from local forms of opera, water sleeves can reach more than two metres in length. It is believed by some that water sleeves are derived from sleeves of Ming dynasty robes, although this has not been proven. The expressive function of water sleeves has developed from being simple

❶ 阙艳华，董新颖，王娜，等. 中英文对照京剧服饰术语［M］. 覃爱东，陶西雷，校译. 北京：学苑出版社，2017：56.

❷ 同❶56，略有改动.

❸ 同❶50.

❹ 同❶50，略有改动.

❺ 同❶58.

to complex, and it is now regarded as an essential art of performance. By manipulating the sleeves, actors can amplify the characters'gestures or express the complexity and richness of a character's feelings. Water sleeves with different lengths, textures, and style, enable more experienced actors to develop distinct and special uses for the sleeves.❶

译文剖析　此处的译文运用了增译法，协调补充了相关信息，便于读者更好地理解水袖的作用。表演者可以通过使用水袖来做出各种各样的动作以增加形象上的美感和丰富人物的思想感情。

（十五）水裙文化内涵及其协调翻译

原文　水裙是一种双层白布短裙，由白色春绸制成，使用时系于腰间。主要为渔夫、樵夫、店小二穿用。❷

译文　Shuiqun (Water Skirt) is a short skirt made of white cloth. It is mostly worn by fishermen, woodmen and waiters. It usually goes with the Chayi (common clothes for laboring people) costume.❷

译文剖析　此处译文补充了水裙的固定搭配茶衣，协调译文与原文的差异，方便读者理解该套服饰的穿用群体。

　　本章小结　｜　本节选取大衣箱中颇具代表性的服饰，具体包括蟒、官衣、开氅、鹤氅、帔、八卦衣、法衣、僧衣、褶子、宫装、古装、裙、袄、裤、旗袍、富贵衣，以及大衣箱中的主要配饰（坎肩、斗篷、蓑衣、领衣、饭单、四喜带、偏带、玉带、丝绦、腰巾、腰箍、云肩、水袖和水裙）等，采用英汉文本并列的方式，从不同角度关注文化差异，并灵活运用不同翻译策略协调英汉语言文化差异，最大程度地忠实于原文文化内涵，补充原文的"弦外之音"，并重视京剧服饰的美学表达，传递文化内涵，促进京剧文化的交流与繁荣发展。

❶ 阙艳华，董新颖，王娜，等. 中英文对照京剧服饰术语［M］. 覃爱东，陶西雷，校译. 北京：学苑出版社，2017：58，略有改动.

❷ 同❶57，略有改动.

第四章

京剧服饰二衣箱文化阐释
及翻译协调美

第一节
京剧二衣箱服装文化内涵及其协调翻译

（一）箭衣文化内涵及其协调翻译

箭衣是满族旗人骑马射箭、打猎时穿着的一种长袍型服装，箭衣样式为里面衬有护领的小圆领，紧身束腰，大襟窄袖，袖口是带有可遮住手的方便射箭的马蹄形袖或敞袖，衣长至靴，自腰际以下左右前后（四面）开衩。

箭衣分为龙箭衣、花箭衣和素箭衣三种，颜色丰富，用途较广，上自皇帝、高级将领，下至一般武士、绿林人物以及衙役等均可穿用。龙箭衣身上绣云龙海水，大将、驸马等身份尊贵的人穿龙箭衣；花箭衣身上绣团花图案；素箭衣为素色无花图案，衙役等身份较低的人穿素箭衣。箭衣根据其材质又可分为缎制和布制两种。穿箭衣时需系鸾带，穿龙箭、花箭时需加三尖领，箭衣也可以作为靠的衬衣用。中外学者对此有如下描述。

原文 Originally, a Mandarin jacket for hunting or fighting, the archer's dress(Jianyi), was first selected for the stage by a Kunqu troupe in south China during the reign of Emperor Qianlong (1736–1796). But it did not come into use for Peking Opera until about a century later, when Cheng Changgeng, a troupe leader as well as a costume designer, used the suit in an altered form. Another Mandarin jacket, habitually labeled a horseback riding suit or Magua, was also adapted to theatrical use at the same time. Taken as a sort of informal attire worn by actors portraying a person in retirement or relaxation after, for example, trudging wearily, these jackets are worn by many characters anachronistically.

The archer's dress, it was quite distinct from that traditionally worn by people in central China, strikingly features horses-hoof-shaped sleeves and four openings in the lower part of this exotic-looking tribal jacket. The sleeves are tightly fitting and the openings are loosely symmetrical: all of these features are supposed to enhance quick shooting and daring horseback riding. Like most ordinary costumes, the archer's dress has a round collar and a lapel, and comes in both primary and the secondary colors. There are three main types: the dragon-decorated, flower-decorated and plain.❶

拙译 箭衣源于清朝满族旗人骑马射箭、打猎时穿的服装。乾隆皇帝统治时期（1736—1796）由南方的昆曲剧团作为戏服搬上舞台。直到大约一个世纪之后，剧团团长、服装设计师程长庚先生对箭衣进行了改造，由此开始在京剧中使用。另一种中式褂子，被习惯性地称作是骑马套装或马褂，同时也适用于戏剧表演，主要用于刻画剧中人物在经过长途跋涉之后，告老还乡或呈放松态的非正式服装。因为此时如果穿戴箭衣就显得极为不合时宜。

作为弓箭手的服装，箭衣有别于我国中原地区民众的传统着装。这种充满满族情调的箭衣，其突出特点在于马蹄袖和衣服下部的四处开衩。袖子很紧但是开口处松散对称，箭衣的所有这些功能都能实现骑马灵活和快速射击。像大多数普通服装一样，箭衣样式为内衬有护领的小圆领，并且有主色和副色。箭衣款式主要有三种——龙箭衣、花箭衣和素箭衣。

原文 Archer's uniforms, are a derivative of dragon robes popular in the Qing dynasty (1644–1911). The Manchus, who came to rule China in 1644, excelled in horsemanship and archery. For men, Jianyi were both formal dress and battle or hunting outfits. In Peking Opera performances, Jianyi has a round opening at the top, and inside the opening there is a high, pure white collar. Jianyi has tight sleeves and each has a cuff in the shape of a horse hoof. The long uniform covers the insteps, but it is tight with a belt around the waist. There are three types of Jianyi—those embroiderd with dragons, for generals and emperors' son-in-laws or brothers-in-laws; those embroidered with flowers, meant for people of middling status; and those without any embroidery, for people like soldiers and bailiffs in government offices.❷

拙译 箭衣是清朝（1644—1911）流行的龙袍的衍生物，满族人因擅长骑马和射箭，于1644年开始统治中原。对于男性而言，箭衣既是正式服装也是作战或狩猎的服装。在京剧表演中，箭衣上部有一个圆形的开口，里面是纯白色的高领。箭衣有收口的紧袖，且每个袖口都呈马蹄形。长的戎装覆盖脚背，腰部用腰带扎紧。箭衣有三种类型，穿戴龙纹刺绣的是大将和皇帝的女婿或姻兄姻弟；那些绣有花卉的，适合中等阶层；而那些没有任何刺绣图案的箭衣，适合官府衙役和执法官员。

原文 Jianyi is influenced by Manchu. It has tapering sleeves, rounded neckline, and rounded crossover front flap closed on the right with toggles and loops. Jianyi was belted for ease of movement. The narrow sleeves ending in horseshoe-shaped cuffs were designed to protect the

---

❶ PAN XIAFENG. The Stagecraft of Peking Opera [M]. Beijing: New World Press, 1995: 123, 略有改动.
❷ 赵少华. 中国京剧服饰 [M]. 北京: 五洲传播出版社, 2004: 54.

wearer's hands when riding. Cut without a shoulder seam, a separate matching collar is worn around the neck of the highest-status characters. Jianyi comes in a full range of colors, with their standard meanings.❶

拙译 箭衣受满族的影响，有着锥形袖子，圆形交领且左压右，并用纽扣与纽襻使其固定。箭衣收腰的穿着形式使运动更为舒适；马蹄袖的设计用于保护骑行者的手；没有肩缝的剪裁和独立的护领是最高级别角色的象征。此外，箭衣颜色丰富，内涵各不相同。

1. 素箭衣文化内涵及其协调翻译

素箭作为箭衣的一种，以缎或绒料制作，此衣之主体部分色彩素净单一，衣无纹样，仅在服装边缘与马蹄袖的袖口增加与主体色彩不同的宽边布料装饰，颜色一般为蓝色或黑色，属于身份较低的人物穿用。素箭衣在穿用时需要搭配靠，即作为衬衣穿在靠内。

原文 素箭衣简称"素箭"。由缎料和布料制成，素面无绣，领口和开衩等处镶异色宽边。常见的服色有红、黑、白、蓝、紫、灰等。主要为公差、衙役、老军、兵士穿用，如《战马超》中的马超、《走麦城》中的关平。❷

译文 Sujianyi (Plain Archer's Uniform) is made of silk or cloth in black, blue, purple, white or grey. By "plain" it is simply meant that there is no embroidery. Decoration comes from the double hems around collar, the side opening and side slits, which are blue. This costume is worn by court runners or soldiers, such as Ma Chao in *Bat against Ma Chao* (*Zhan Machao*), and Guan Ping in *Loss of Maicheng City* (*Zou Mai Cheng*).❷

原文 The Sujianyi is made of satin and comes in black, white, gray and blue. Contrasting borders highlight the edges of the robe. Lower assistants wear this.❸

拙译 素箭衣是由黑色、白色、灰色和蓝色的缎子缝制而成，利用与袍服主体颜色形成对比色的镶边来提亮整个素箭衣，一般为身份低下的侍者穿用。

2. 团花箭衣文化内涵及其协调翻译

团花箭衣是指箭衣上所绣的图案为团花或花枝，有的甚至在衣服的边缘也绣有相

❶ BONDS ALEXANDRA B. Beijing Opera Costumes: the Visual Communication of Character and Culture [M]. Honolulu: University of Hawaii Press, 2008: 151.
❷ 阙艳华，董新颖，王娜，等. 中英文对照京剧服饰术语 [M]. 覃爱东，陶西雷，校译. 北京: 学苑出版社，2017: 72, 略有改动.
❸ 同❶153–154.

对应的图案。其相关英文与汉译描述如下。

原文 Huajianyi is a floral uniform in various colours, with eight embroidered rounded flower or rounded bat designs with decorative figures or stripes around the collars and sleeves. It is designed for an army officer, an outlaw or a young warrior.[1]

拙译 团花箭衣上绣有八个团花或八团蝙蝠，领口和袖口有图案或条纹装饰。该设计适用于武艺高强的武将、绿林豪杰或少年英雄。

原文 The Huajianyi has flower roundels and a border around the edges or simply a flower border. Fighters always wear this.[2]

拙译 花箭衣绣有团花图案和镶边装饰，或为简单的花边装饰，常为武士穿用。

3. 龙箭衣文化内涵及其协调翻译

龙箭衣是箭衣的一种，因其衣上所绣的图案为团龙、行龙或蟒的图案而得名，有红、黄、黑、白、绿、粉等不同色彩。其相关英文与汉译表达如下。

原文 The dragon decoration implies royal attire in accordance with Chinese tradition. Designs of both legendary animals and waves perhaps suggest more activity than sedateness, so it has been typically displayed by Liu Bei, one of the pretenders to the throne during the Three Kingdoms Period, in *Return to Jingzhou*, and by Xue Pinggui in *Wujia Slop*. A defeated general also wears one as Hua Yun does in *Battle at Taiping*.[3]

拙译 龙图案的装饰与中国传统中皇室的装束一致。传说中动物和波浪的设计或许暗示着更多动态与活跃，而这一点通过三国时期的刘备，在京剧《回荆州》时给予了特色传达，而京剧《武家坡》中的薛平贵也同样如此，甚至在《战太平》中，败将花云也身着龙箭衣。

原文 Longjianyi, decorated with dragons in colors for standard males or couched in gold thread for Jing, is worn by emperors and other high officials connected with the court.[4]

拙译 龙箭衣采用不同色彩的龙的图案作为装饰，一般为成年男子穿用；或用金线绣制为京剧中的净角穿用，适合皇帝和其他宫廷高官的扮演者。

---

[1] 阙艳华，董新颖，王娜，等. 中英文对照京剧服饰术语［M］. 覃爱东，陶西雷，校译. 北京：学苑出版社，2017：70，略有改动。

[2] BONDS ALEXANDRA B. Beijing Opera Costumes: the Visual Communication of Character and Culture［M］. Honolulu: University of Hawaii Press，2008：153.

[3] PAN XIAFENG. The Stagecraft of Peking Opera［M］. Beijing: New World Press，1995：123-124.

[4] 同[2]152-153.

（二）靠文化内涵及其协调翻译

靠又称甲，源于清代将官之绵甲戎服，上衣下裳相连，具有长宽袍的庄重大方。圆领，紧袖，长及足。各种正色、副色具备。材质为各色素缎子，素缎上又用金、银、彩线绣满了鳞甲，对色彩的应用与蟒相同。靠身（甲裳）分前后两片，其上部、下部及两肩绣鱼鳞纹或丁字纹；中部靠肚略阔，长方形，硬里、凸起，绣虎头或龙纹；另有护腿两片称靠腿或下甲，系于腰部左右。靠的结构复杂，全身共有绣片31块，其中有3块可移为他用。穿蟒或箭衣者围靠领，象征武将；单用两块靠腿者，象征丢盔卸甲的败将，具有符号意义。同时，靠又有软、硬之分，硬靠需背三角形靠旗4面，插靠旗的硬皮壳称背壶；软靠不背靠旗，用于非战斗场合的武将。女靠与男靠大致相同，但女靠的靠肚以下缀有飘带。此外，穿男靠加三尖领，女靠加云肩。小生及不戴髯口的武生扎靠时，胸前有的需加彩球，而女靠胸前有的加护心镜。另有霸王靠，黑色，靠肚下沿缀有黄色排穗。霸王靠为西楚霸王项羽专用，黑色平金绣，其最显著的特点是靠肚下端缀有黄色的网子穗。而清代宫廷演剧的霸王靠绣象鼻，甲片为方形，但现已被淘汰。类似英文与汉译描述如下。

原文 The armor, or Kao, for soldiers on the stage is, in the main, divided into "hard" and "soft" types. The hard armor has four pennons attached to the back of the wearer while the soft does not. In structure the armor is, in fact, an assemblage of four single parts in addition to a collar that is reinforced and in disc shape, and a pair of sleeves that are tapering and tied up at the ends. The front piece and the back piece are both decorated with fish-scale patterns. Perhaps with a view to alluding to the sense of defense or even invulnerability, a shield-like protection piece (Kaoduzi) with dragon or tiger's head patterns is closely fastened to the front, and a leather holder for pennons is fixed on the back. On both the left and right, dangle pieces of the so-called lower armor (Xiajia) or rather side pieces (Kaopaizi) to protect the thighs.❶

拙译 靠是京剧舞台上的士兵的铠甲，主要分为硬靠和软靠两种类型。硬靠有四面细长的三角旗插在穿衣者的背部，而软靠则没有。在结构上，除了领子外，铠甲实际由四个独立部分组成，呈片状，加强了防护作用。两只袖子呈锥形，在袖口处扎起。前片和后片均装饰有鱼鳞花纹。也许是为了暗指防卫意识或坚不可摧的精神力量，一种带有龙或虎头图案的盾状保护片（靠肚子）紧紧地固定在前胸，后背上固定着皮袋用来插细长的三角旗。左右两侧悬挂所谓的下护甲（下甲）或侧件（靠牌子）以保护大腿。

原文 Kao, as a kind of decorative armor, falls into three categories: hard, soft and modified.

❶ PAN XIAFENG. The Stagecraft of Peking Opera [M]. Beijing: New World Press, 1995: 120.

114

The hard armor consists of a leather sheath tied to the back of the actor and four embroidered triangular flags, this makes them all the more imposing and commanding.❶

拙译 靠作为作战时战士穿的铠甲可分为硬靠、软靠和改良靠。靠背后绑着一个插着四面三角形靠旗的皮鞘，靠旗源于古代将官的令旗。传令时执于手中，骑马行进时就插在后腰革带上。在京剧中靠以靠旗为装饰物，演员穿大靠时会衬四面靠旗，看上去特别威猛神勇。

原文 Kao comes from the ancient warriors armor which was developed after evolution through decoration and beautification. Kao is a kind of Military Costumes for Military Generals and certain female General the war occasions. Kao has two styles for male and female. Kao for male is decorated with patterns of fish-scale or three-way scale as armor. When the actors wear Kao, they must wear Kaoqi at the back to indicate that the role is already fully armed. It is called as "Yingkao" or "Dakao". If wearing Kao but without Kaoqi, it indicates that the role is not at the war occassion and it is called as "Ruankao".❷

拙译 靠源于清代上衣下裳制的戏服，起到美化与修饰的作用。靠是战争中将军或某些女将的戎装。款式分为男、女两款。女靠用鱼鳞或丁字纹作为铠甲的装饰。男演员穿靠时，背后插着靠旗体现其全副武装的状态，被称作是硬靠或大靠。如果只穿靠，没有靠旗，就意味着不是战争状态，被称作软靠。

原文 Decorative armor with images of fierce beasts and complex pieces has bedecked Chinese generals for years. High-ranking generals and military officers wear the Kao. The front of the Kao has a wide padded section at the waist (Kaodu), the waist piece in the back is soft, conforming to the contour of the body; it is tied around the hips and connects to the front piece with tapes. Below the front and back waist are two layers of rounded aprons, the upper one reaching to the knees and the lower one to ankles. This tabard unit also has sleeve pieces attached, with larger flaps that cover the shoulders and upper arm sewn together as a single unit. The sleeves are set into upper half of the armscye along with the shoulder flaps, but they are left open under the arm and only tacked together below the elbow. Beneath this main piece, leg flags (Kaotui) resembling chaps are worn. The four armor flags (Kaoqi) are mounted on a stand with four poles and lashed to the actor's back. When the flags are strapped onto the back, the garment becomes the Yingkao, which is worn by military officers. When the flags are not worn, the armor is called Ruankao. The Kao comes in most of the upper and lower colors. The centers of the front and back pieces on the chest and the aprons are embroidered with either the fish-scale design or an interlocking three-pronged geometric

❶ 赵少华. 中国京剧服饰 [M]. 北京：五洲传播出版社，2004：66.
❷ 谭元杰. 中国京剧服装图谱 [M]. 北京：北京工艺美术出版社，2008：77.

shape. The Kao is worn with the padded vest underneath to increase the scale and presence of the figure.❶

拙译 多年来，我国军服都绣有猛兽和复杂图案。高级将领和军官穿靠。靠的前片腰部有宽大的衬垫（靠肚），后片的软腰柔软贴体，系于臀部附近，与前片用两根搂带连接（腋下的燕窝也起到了连接前后片的作用）。前后腰的下部有两层，上面的一层长及膝盖，下面的长及脚踝。而且靠的前臂、盖住肩膀的苫肩和上臂部分缝制在一起，即靠的袖子和苫肩上半部分缝合，手臂下部开口，仅在肘部以下的地方缝缀起来。绣有类似甲片纹样的、能够遮护腿部的靠腿穿在靠的前后片内。四面靠旗插在演员的后背，这种铠甲称作硬靠；当不插旗子的时候，铠甲称作软靠。靠的颜色多数是上五色和下五色。前后片护胸的正中和下部带装饰，有鱼鳞形状的甲片纹样或三叉交错的几何图形（丁字纹）。靠的下面借助垫肩，塑造出武将高大威猛的形象。

1. 男靠文化内涵及其协调翻译

原文 "男靠"又称"大靠"。其结构非常复杂，由靠身、下甲、靠领、靠旗等部件组成。全身共有31块绣片，其中有3块可以移为他用。穿蟒或穿箭衣者围靠领，象征武将；单用两块靠腿者，象征丢盔卸甲的败将，具有符号意义。武将身份和性格的差异在靠的颜色上也可以看出来。一般来说，红靠为元帅，绿靠为大将，黄靠为贵族、老将，黑靠为彪悍猛将，白靠为年轻武将。❷

译文 A whole set of Nankao (hard armour / armour for men) is made up of 31 embroidered pieces, three of which can be removed for other purposes. A ceremonial robe called Mang or archers uniform called Jianyi with an armour collar would suggest a military officer, whereas two legs shields would mean a defeated general fleeing skelter. Different colours indicate different status or personalities. Generally, red armours are worn by the commander-in-chiet, green ones by generals, yellow ones by noble persons or veteran officers, white ones by young officers and black ones by valiant officers.❸

2. 女靠文化内涵及其协调翻译

女靠的服装造型比男式靠更具有装饰性，色彩纹样都更为绚丽，完美地衬托出女将英武和阴柔之美的形象气质，尤其是女硬靠，其靠肚略小，绣双凤牡丹。靠肚下缀两层或三层飘带。虎头肩（肩部甲片）下另衬荷叶袖，也绣凤及牡丹，用时加用云肩。

❶ BONDS ALEXANDRA B. Beijing Opera Costumes: the Visual Communication of Character and Culture ［M］. Honolulu: University of Hawaii Press, 2008: 127–129.
❷ 阚艳华, 董新颖, 王娜, 等. 中英文对照京剧服饰术语 ［M］. 覃爱东, 陶西雷, 校译. 北京: 学苑出版社, 2017: 63.
❸ 同❷63, 略有改动.

其英文表述如下。

原文 Hard armor for female generals or warriors is characterized by a piece tied around her waist embroidered with phoenix on either side of a peony flower. Note the cape on the shoulder.❶

拙译 女靠比男靠更具装饰性。女靠的靠肚略小，靠肚下缀彩色的飘带。肩部下衬荷叶袖，绣凤及牡丹。穿时配云肩。

原文 Kao for female is decorated with patterns of phoenix and peony which are particularly beautiful and gorgeous.❷

拙译 女靠与男靠大致相同，唯靠肚稍小，腰下有彩色飘带。用时围云肩，系衬裙。

原文 The Nükao resembles the men's version of armor, and has a narrow padded belt, two additional layers of multicolored pastel streamers in the skirt era. It is worn with a more elaborated, larger cloud collar. Nükao are most commonly constructed in red and pink, and the ornamentation are multicolored phoenixes, peonies and flowers.❸

拙译 女靠类似于男靠，但有窄腰带，裙边还饰有两层彩色飘带，穿时配有更精致和更大的云肩。女靠最常见的颜色是红色和粉红色，并装饰有五彩凤凰、牡丹和花朵等图案。

3. 霸王靠文化内涵及其协调翻译

原文 "霸王靠"样式与男靠相同，为西楚霸王项羽专用。靠身为黑色平金绣，靠肚下端缀有黄色排穗。❹

译文 Bawangkao (Hegemon's Armour) is used exclusively for the Hegemon King of Chu, Xiang Yu. In history, Xiang Yu and his rival, Liu Bang, fought with each other for the domination of China. What is special about this armou is that it has tassels on the front.❺

译文剖析 霸王靠为西楚霸王项羽专属，译文补充了相关的历史背景，便于读者理解霸王靠的特殊性。基于原文，译文可尝试补充"it is black with flat golden emboridery"，为了凸显该靠肚下的黄色排穗，以区别于其他男靠，不妨在"tassels"前补加"yellow"，给予协调修饰。

---

❶ 赵少华. 中国京剧服饰［M］. 北京：五洲传播出版社，2004：70.
❷ 谭元杰. 中国京剧服装图谱［M］. 北京：北京工艺美术出版社，2008：77.
❸ BONDS ALEXANDRA B. Beijing Opera Costumes: the Visual Communication of Character and Culture［M］. Honolulu: University of Hawaii Press, 2008：131.
❹ 阚艳华，董新颖，王娜，等. 中英文对照京剧服饰术语［M］. 覃爱东，陶西雷，校译. 北京：学苑出版社，2017：66.
❺ 同❹66，略有改动.

4. 改良靠文化内涵及其协调翻译

改良靠是京剧大师周信芳所创，将靠改为上下两部分，用束腰使其紧身合体。靠腿分前后左右共四块，软带上及肩部有半立体虎头。甲片缀排穗，不扎靠旗。这种服装造型简洁轻便。有的在改良靠上不绣鳞甲，只绣上花纹或镶上金属凸片。改良靠不及传统靠的样式威武，一般只用于普通将官。其相关的英文及汉译表述如下。

***

原文 The Gailiangkao was designed by the actor Zhou Xinfang. The shapes of the layered pieces are similar to those of the Kao, retaining the look of defense, while slimming and streaming the form to facilitate movement in martial arts scenes. The key distinctions are the absence of the four flags on the back and the padded belt at the front waist. The jacket has either a center-front opening or a crossover flap to the right side and shares the shoulder, armscye, and sleeve construction of the Kao. The peplum, cut without a waist seam, can either be a part of the decorative layering or plain and covered by the leg flaps. The greatest deviations in shape occur below the waist there the panels may take on various profiles. The lower sections generally include two or three full-length leg flaps on the sides and back, and shorter rounded layers in the front and back.These pieces are mounted on two waistbands that tie around the waist.❶

拙译 改良靠是由京剧演员周信芳设计。分层的造型类似于其他的靠，同时保留了防护的外观，在造型上更加简洁轻便，适合舞台上动作表演的展示。改良靠的关键在于省去了背部的四个护旗和前腰的衬垫。改良靠要么是前开襟、要么是右开襟，而且在肩部、袖窿和袖子的结构上和普通靠一样。靠腿覆盖的腰裙没有腰缝，是一层有装饰或者素色的面料。改良靠在外形上最大的改变是腰部以下的衣片形状变化多样。一般腰部以下由两层衣物组成，下面一层包括两片或三片长的下甲，用腰带系在腰间，垂在大腿的两侧和身后；上面的一层是较短的圆形的下甲，挡在腿部前后。

（三）抱衣抱裤文化内涵及其协调翻译

抱衣抱裤分为花抱衣抱裤与素抱衣抱裤两类。花抱衣抱裤以某种颜色的绸缎制作，衣与裤上面都绣有相同的彩色图纹，衣领、下摆及袖口等处则绣花边。穿时腰系鸾带，胸背部系绳，足穿相配的薄底靴。穿花抱衣抱裤的人物有很多，而穿素抱衣抱裤的人则有限，多为有武艺的老者。因其年龄不适合漂亮的色彩与图案，所以素抱衣抱裤一般不绣花，而在下摆、衣领和袖口等处绣有如意头线条，作为简单的装饰，颜色也一般为香色和古铜色。

但仅就抱衣而言，也称豹衣或英雄衣。俗话把衣紧合身称为抱身儿。抱衣的最大

---

❶ BONDS ALEXANDRA B. Beijing Opera Costumes: the Visual Communication of Character and Culture［M］. Honolulu: University of Hawaii Press，2008：133.

特点是紧瘦抱身，裹身窄袖，大领大襟，衣长为半身短衣，衣之下摆处连缀两重绸子走水（旧时为三层，现在常缝缀两层），缎制绣花。类似英文表述如下。

原文 One type of costume for combat performance is widely known as the hero's suit. It consists of both a jacket and a pair of pants, which are often professionally dubbed as Baoyi and Baoku respectively in Peking Opera, that may be literally rendered as "panther's jacket and pants". The jacket has a large collar and lapel, and two layers of pleated white silk fabric on the lower part. The pants are loose and plain as compared with the upper garment, that is tight-fitting and dragon-patterned. The suit is invariably in one and the same color.❶

拙译 京剧中有一款适合打斗表演的服装就是广为人知的英雄衣。英雄衣由上身褂衫和下身长裤组成，在京剧中的专业术语是抱衣和抱裤，按照字面意思可能会理解为豹衣裤。上身褂衫是大领大襟，褂子的底襟处缝着两层丝绸边。裤子与上衣相比更宽松平整，贴身且有龙纹图案。全身采用同一颜色。

原文 Tight-fitting jackets and trousers are also known as "'heroes' outfits". Tight and elegant, they are for chivalrous men. It has a broad collar, tight cuffs and a front that buttons on the right. What is peculiar about it is that silk ribbons in three layers are sewn to the lower part of the hem, making the jacket look like a skirt.❷

拙译 抱衣抱裤为侠客义士等英雄人物穿用。抱衣的样式为大领大襟，紧身紧袖，在褂子的底襟周围缝着三层宽绸边，一层摞着一层，像裙子似的，上面还带有绣花。

原文 "抱衣抱裤"又称"包衣包裤""豹衣豹裤""短打衣"。大襟，大领，紧身，束袖，衣长及臀，下摆加饰走水。走水上用彩线刺绣团花、蝙蝠等图案。抱衣抱裤分花、素两种。花抱衣全身彩绣团花、金钱等图案，一般为绿林好汉穿用，如《红桃山》中的花荣、《李家店》中的黄三太、《失子惊疯》中的金眼豹。素抱衣的缎面无绣，只是在领口、大襟、褂下与袖口等处镶饰各种图形的异色缎边，主要为年迈的江湖人士等穿用，如《嘉兴府》中的鲍赐安、《四杰村》中的花振芳、《打渔杀家》中的萧恩。❸

译文 Baoyi Baoku (Tight-fitting Jackets and Trousers) is a tight fitting suit won by male roles. It has crossover collars and jacket buttons down the front. The cuffs are relatively tight. The hem is decorated with two layers of white silk and look like waves of water, known as Zoushui (running water). The satin is embroidered with patterns of small flowers and the cuffs are embroidered with gold thread. There are two types: one is embroidered with rounded

❶ PAN XIAFENG. The Stagecraft of Peking Opera [M]. Beijing: New World Press, 1995: 124–125.

❷ 赵少华. 中国京剧服饰 [M]. 北京: 五洲传播出版社, 2004: 80.

❸ 阙艳华，董新颖，王娜，等. 中英文对照京剧服饰术语 [M]. 覃爱东，陶西雷，校译. 北京: 学苑出版社, 2017: 82.

京剧服饰二衣箱文化阐释及翻译协调美 | 第四章

patterns and the other is plain without any patterns. The former is embroidered with flowers and shapes derived from old coins, and the cuff is embroidered with golden and multi-coloured threads. Hua Rong in *Red Peach Mountain* (*Hongtaoshan*), Huang Santai in *Lee's Inn* (*Lijiadian*), Jinyanbao in *Going Mad After Losing a Son* (*Shizijingfeng*) all wear this kind of costume. The second type is worn by bandits, mountain raiders, and forest outlaws. Bao Cian in *Jiaxing Mansion* (*Jiaxingfu*), Hua Zhenfang in *Four Heroes Villages* (*Sijiecun*) and Xiao En in *Fisherman's Revenge* (*Dayushajia*) wear this kind of costume.❶

（四）打衣打裤与战裙文化内涵及其协调翻译

打衣打裤是女式英雄衣的专用名词。褂子式样是小领、对襟、紧袖，并有绣花。虽然没有走水，可是它在腰下系着两片下甲，贴靠在大腿的两旁，同时包住臀部，这两片甲片代替了裙子，故又名战裙。因为身着女式英雄衣的人物总是要进行战斗和开打，穿裙子不太方便，但是只穿裤子又略显单调，所以在下身系着战裙。像《武松打店》中的孙二娘和《打焦赞》中的杨排风等女将、女英雄和女强人就穿打衣。其英文概述与汉译如下。

原文 Small collars, buttons down the front and tight cuffs characterize the combat outfits. Such outfits are embroidered, with two piece of armor tied to the thighs.❷

拙译 打衣打裤样式为小领、对襟、紧袖，上有绣花。腰下靠在大腿两旁系着两片下甲，同时包住臀部。

（五）快衣快裤文化内涵及其协调翻译

快衣快裤也称侉衣侉裤，与抱衣抱裤的性质相似，也是紧身的，但其样式是圆领、对襟、紧袖，下摆无走水，多以黑缎制成，一般没有什么绣花，是短打武生和武丑经常穿的短衣服，胸前领口至下摆边以及袖口到衣衩共有三排白色扣襻儿，俗称疙瘩襻儿，也有的在边上加云头，扣襻下身是纯黑色的裤子。穿快衣快裤一般系鸾带。武丑大部分都穿快衣快裤；而武生只有一部分穿，像武松、石秀以及《巴骆和》中的余千等。快衣也分花、素两种。素快衣一般用黑色缎制成，前胸自领口至下摆、两袖口至腋下衣衩缀有密排的白扣襻（称为英雄结），讲究的扣襻用象牙或骨料制成。这三排白扣襻纯属美化服装的装饰，它以三条白色垂直线对大面积的黑色进行分割，造成强烈

---

❶ 阙艳华，董新颖，王娜，等. 中英文对照京剧服饰术语［M］. 覃爱东，陶西雷，校译. 北京：学苑出版社，2017：82，略有改动.

❷ 赵少华. 中国京剧服饰［M］. 北京：五洲传播出版社，2004：82.

的黑白对比，有效地衬托出黑色的俏美潇洒。素快衣通常用于短打武生行当的江湖英雄或反面人物，也用于家丁等群众角色，是一种适应面较广的武服。其相关的英文与汉译表述如下。

原文 Closely related to the hero's suit for combat scenes may have been the military blazer or sports jacket. In Chinese, it is conventionally called Kuaiyi. It is usually coal-black but adorned with white fasteners in three rows, of which one runs down the breast while each of the other two range from the shoulder to the end of the sleeve, which is close-fitting against a round collar and a large lapel. Considered one of the best costumes for acrobatic feats, it is also clown's garb.❶

拙译 京剧中与英雄的战斗场景密切相关的服饰是武生的短衣，通常称作快衣。快衣的颜色为煤黑色，由三排白扣襻装饰，其中一排在前胸自领口至下摆处，而另外两排自两袖口延伸至腋下衣衩，与圆领和大翻领紧密贴合。快衣是武打表演中最好的服装之一，也是武丑的重要装束。

原文 The most common Kuaiyi set is black. As the black color of the garments aids in their disguise in the dark, they are sometimes called clothes for the night. The functional closings are concealed, and decorative frogs in contrasting colors go down the center front, the side seams, and from shoulder to wrist outside the sleeves. Kuaiyi with white decorative frogs, known as heroes' buttons, are worn by warriors and those with gold frogs are worn by Jing. A black velvet Kuaiyi jacket worn with silk trousers and a white or yellow sash is specific to Wu Song, a principal character in *the Water Margin Stories*. The hua Kuaiyi comes in many colors and has designs of birds or butterflies embroidered on it to represent the wearer's agility and strength.❷

拙译 最常见的快衣是黑色。衣服的黑色有助于在黑暗中伪装，因此有时又被称为夜服。快衣的作用在于隐蔽，前胸自领口至下摆、两袖口至腋下衣衩都有装饰性的扣襻，其颜色与衣服形成对比。"快衣"采用这种白色纽扣又被称为英雄扣，为武士穿用；而金色扣襻装饰的快衣则为京剧中的净角穿用。《水浒传》中的主要人物武松，就穿着特制的黑天鹅绒快衣，以及配有白色或黄色丝绦的丝质裤子。"花快衣"有多种颜色，其上绣有鸟或蝴蝶图案，暗示了穿衣者的敏捷性与力量。

原文 The Bingyi (soldier's clothes) version of the Kuaiyi costume is worn by soldiers in the armies of the principal generals. The design on these garments can be a simple circular medallion on the chest and borders on the hem, cuffs and neck, or a more elaborate scattered

❶ PAN XIAFENG. The Stagecraft of Peking Opera [M]. Beijing: New World Press, 1995: 125.
❷ BONDS ALEXANDRA B. Beijing Opera Costumes: the Visual Communication of Character and Culture [M]. Honolulu: University of Hawaii Press, 2008: 163.

composition of animals, mythological creatures, or lucky symbols. The color of the Bingyi relates to the color of the leader. Two opposing armies generally wear different colors of the same style Bingyi. Both groups will have either the simpler medallion decoration or the version with more detailed embroidery. Immortal soldiers may have lucky charms embroidered on their costumes.

All styles of the Kuaiyi are worn with a white collar inside the neckline. One of several belts is used to constrict and secure the waist: a stiffened fabric belt matching the jacket, a sash, or a firm belt with tasseled ends. The excess fabric of the jacket body is pulled into a box pleat in the back to create a smooth line in the front.❶

拙译 快衣款式的兵衣（士兵服）为大将麾下的士兵穿用。这些服装的设计有的是胸前带有简单的圆形图案，下摆、袖口和领口上有边饰；有的是打散的动物、神话故事形象或吉祥纹样的组合。兵衣的颜色与将领的服色有关。两支对立的军队通常穿着不同颜色、但是同样风格的兵衣。每支军队的服装都有更简洁的图案装饰或更关注细节的刺绣。如天兵天将的服装上可能都绣有幸运符。

所有款式的快衣都配有白领，还有好几条带子，其中一条腰带用于收紧和固定腰部。扎紧的带子、腰带或尾端带流苏的腰带一般与上衣相配，可以将上衣多余的部分塞入背部的方褶中，方便前面整理出平滑的效果。

原文 "侉衣侉裤"又称"快衣快裤"。与抱衣抱裤的性质相似，是紧身的，便于武打。圆领，对襟，束腰，下摆无走水。分花、素两种。"素侉衣侉裤"多以黑缎制成，全套衣服黑白分明，主要用于江湖英雄或反面人物，也用于家丁等群众人物，如《白水滩》中的穆玉玑、《四杰村》中的余千、《清风寨》中的李逵等。"花侉衣侉裤"的使用者较少。其通常由黑色丝绸制面，周身彩绣素色飞蝶或飞蝠图案，主要为动作敏捷的武丑穿用，如《花蝴蝶》中的蒋平、《连环套》中的朱光祖、《九龙杯》中的杨香武等。❷

译文 Kuayi Kuaku (Embroidered Combat Outfit) is a light weight combat outfit made of black satin. It has a small collar, with a right-buttoning front and tight sleeves. The trousers are all in black. Mu Yuji in *The White Water Beach* (*Baishuitan*), Yu Qian in *Four Heroes Village* (*Sijiecun*), Li Kui in *Qingfeng Village* (*Qingfengzhai*) wear this outfit. Embroidered Kuayi Kuaku is seldom used and is almost identical to the plain black one. The black satin is usually embroidered with flying butterfies or flying bats and is usually worn by swift acrobatic fighting clown roles such as Jiang Ping in *The Butterfly* (*Huahudie*), Zhu Guangzu in *A Strategem of Interlocking Rings* (*Lianhuantao*), and Yang

---

❶ BONDS ALEXANDRA B. Beijing Opera Costumes: the Visual Communication of Character and Culture [M]. Honolulu: University of Hawaii Press, 2008: 163-164.

❷ 阙艳华, 董新颖, 王娜, 等. 中英文对照京剧服饰术语 [M]. 覃爱东, 陶西雷, 校译. 北京: 学苑出版社, 2017: 81, 略有改动.

Xiangwu in *Nine Dragon Cup* (*Jiulongbei*). It is also worn by warriors, heroes and private generals.❶

（六）马褂文化内涵及其协调翻译

　　马褂虽然是清朝的服装，但在京剧里可以出现在任何朝代。颜色分红、绿、黄、白、黑五种。样式和中国近代的长袍外边套的马褂类似，对襟无领，衣长66厘米有余；只是外部加了加工，绣了一些式样为圆形的花纹图案，黄马褂和黑马褂上都绣有团龙与海水的图案；其他颜色的马褂绣龙，不绣海水。穿马褂需内穿箭衣。其相关英文与汉译描述如下。

　　原文 The Mandarin vest, worn over a gown, was popular in the Qing dynasty. In Peking Opera, however, it is for characters of all times, worn as an outer garment over an embroidered combat outfit.❷

　　拙译 马褂源于清朝，在京剧中可为任何朝代的角色穿着。作为一种外罩性质的服装，穿着时套在箭衣外边。

　　原文 马褂。按照清代的服装美化而成。圆领，对襟，无袖，衣长二尺。由缎料制面，周身彩绣团海水图案的为帝王将帅穿用，彩绣团寿或龙纹图案的为中军、旗牌穿用。服色有红、绿、黄、白、黑几种。素面无绣的马褂，主要为身份更低的校尉、随从穿用。使用时，将其套在箭衣外，可作为行路时的外褂。如《武家坡》中的薛平贵、《四郎探母》中的杨延辉。

　　马褂也可以作为战斗时穿的服装，但穿马褂的人物并无激烈的开打场面，而是点到为止，如《碰碑》中的杨六郎。此外，马褂还有其他穿法，如斜穿式，偏袒右肩臂，突出人物急行赶路时的心情，如《罗成叫关》中的罗成和《铡美案》中的韩琪。❸

　　译文 Magua (Manchu Jacket) is around 66 cm in length with a round collar, front opening and loose sleeves. If the costume is embroidered with dragons and sea water, it is worn by emperors, kings, military officials. If it is embroidered with the Chinese character "shou" (meaning longevity), it is used by lower ranking military officials and warriors. Various colours such as red, green, yellow, white and black are used to distinguish people according to their social status. If the garment is without any embroidery, it is for people of

❶ 阙艳华，董新颖，王娜，等. 中英文对照京剧服饰术语［M］. 覃爱东，陶西雷，校译. 北京：学苑出版社，2017：81，略有改动.
❷ 赵少华. 中国京剧服饰［M］. 北京：五洲传播出版社，2004：84.
❸ 同❶76.

humble origion. The Manchu Jacket, worn over a gown, was popular in the Qing dynasty. In Peking Opera performance, however, it is for characters of all historical periods, and worn as an outer garnment over an embroidered combat outfit. Xue Pinggui in *Wujia Slope* (*Wujiapo*), Yang Yanhui in *Yang Visits his Mother* (*Silangtanmu*) wear this gown.

The Manchu Jacket can also be used in battle. The person who wears it does not need to appear in an intense battle scene but only to a point. Such as Yang Liulang in *The Tombstone* (*Pengbei*). There are also other ways to wear a Manchu Jacket, which partially bare the right shoulder and arm. This highlights that the character is in a hurry on his journey. Examples of this type include Luo Cheng in *General Luo Cheng at the City Gate* (*Luochengjiaoguan*), Han Qi in *Executing Chen Shimei* (*Zhamei'an*).❶

（七）猴甲文化内涵及其协调翻译

猴甲是京剧《大闹天宫》中美猴王的铠甲，属于特色服饰，其英文表述与汉译如下。

原文 The rebellious Monkey King is one of the most famous and popular characters of Chinese literature, featuring prominently in the novel "*Journey to the West*".❷

拙译 猴甲作为铠甲的一种，用于传统神话剧《大闹天宫》中的孙悟空穿着。

原文 猴甲是铠甲的一种，用于传统神话剧《大闹天宫》中的孙悟空。❸

译文 Houjia (Monkey King's Armour) is worm exclusively by the Monkey King, one of the most popular characters in the classical Chinese novel *Journey to the West*.❸

（八）卒坎肩文化内涵及其协调翻译

卒坎肩衣长如短衣，圆领，对襟，左右开衩，用红色绸缎料制成，老式卒坎肩在前胸及后背处缀有白色圆形图案，上面墨绣"卒"或"兵"字，是一种装饰在交战双方士兵服装上的标志。近代往往金绣团花和缘饰纹样，以求得更为美观。

原文 For ordinary soldiers there is a sort of sleeveless jacket that may be dubbed the solider's

---

❶ 阙艳华，董新颖，王娜，等. 中英文对照京剧服饰术语［M］. 覃爱东，陶西雷，校译. 北京：学苑出版社，2017：76，略有改动.

❷ 赵少华. 中国京剧服饰［M］. 北京：五洲传播出版社，2004：76.

❸ 同❶75.

jerkin or Zukanjian. Mainly in red or green, it is worn by the scout in *Stealing the Imperial Horse*, the old soldiers in *Fall of Jieting*, and the watchmen in *King's Parting with His Favorite*.❶

> **拙译** 对于普通的武士，有一种无袖的外套被称为兵卒的短上衣或卒坎肩，颜色主要为红或绿，如《盗御马》中的探子、《失街亭》中的老卒，以及《霸王别姬》中的守夜人，无一例外都身着卒坎肩。

（九）龙套文化内涵及其协调翻译

龙套是对襟大袖，有水袖，后身开衩，开衩的接合处以双拼如意头为饰。衣摆底边的贴边稍曲，呈波状。此外，领口、袖口、前身襟边以及后身衩边均镶不同颜色的边，并配以古钱、回纹、草龙图案。袖子及前后身绣有正团龙、侧团龙若干条，配以云头图纹。其底边还绣有少许海水、海浪。龙套四件为一堂，代表卫队。作战时，手执标帜旗，即代表千军万马。龙套颜色一般要与头面角色所着的蟒或靠配套，红色的即称红龙套，白色的即称白龙套。青袍是龙套中的一种，黑色，布制，无水袖。这种形制在褶类中规格最低，用于知县大堂上的衙役。相关的英文表述与汉译如下。

> **原文** Walk-on roles in Peking Opera are generally distinguished by their special garbs, known as Longtao, which have a circular collar, a lapel and a pair of long sleeves. Conventionally in red, green, white or blue, they are decorated with patterns of dragons over the entire surface.❷

> **拙译** 京剧中跑龙套的角色因其特制的衣裙——龙套，而有别于其他扮角儿。龙套有着领口、翻领和长袖。通常是红、绿、白或蓝色，通身上下都有龙的图纹作为装饰。

（十）大铠文化内涵及其协调翻译

大铠用于金殿上的御林军，形制与靠相似，也有靠肚，但没有靠腿子。靠肚以下的甲片分为左右两块。大铠色一般为紫红色，平金绣纹样，使用时不插靠旗。其英文与汉译描述如下。

> **原文** Members of the imperial guards dress in red uniforms akin to armor, with no pennons and no side pieces. But the protection piece, unlike that with the armor, which dangles freely, is made immovable by fixing it onto the uniform, termed in Chinese Dakai. This suit ought to be

---

❶ PAN XIAFENG. The Stagecraft of Peking Opera［M］. Beijing：New World Press，1995：127.
❷ 同❶119.

matched with a special spear known as Hebaoqiang.❶

拙译 皇家御林军一般穿着类似于铠甲的红色制服，没有靠旗和甲片。但是，该制服的防护装置又不同于铠甲，呈自由悬挂状，通过将其固定在制服（大铠）上，方便移动。这套衣服配有特殊的枪，即荷包枪（京剧道具，又称开门枪、银枪头、金合抱枪结，朱红杆较粗长，银钻，多用于御林军，四杆为一堂）。

（十一）茶衣文化内涵及其协调翻译

茶衣是半身的蓝布对襟短衣，一般分为两种：一种是大多数人穿用的斜领、大襟的褶子；另一种是少数人，特别是儿童穿用的对襟褶子。茶衣的袖子有带水袖的，也有不带水袖的。有的角色上边穿着茶衣，下边穿着裤子；也有的是在裤子外边再系一条短裙，这种短裙也叫腰包。凡属特殊身份者，一般都系腰包，如酒保和手艺工匠。由于茶衣色彩单一，样式朴素，除儿童扮演者穿用以外，更多的是一些酒保、樵夫和渔夫等角色扮演者等穿用，所以茶衣实际上是劳动人民和下层人身份的象征。其英文与汉译描述如下。

原文 For children of plebeian origin, we find jackets fastened down the front and known as Chayi. They are worn by the players of such roles as Xue Yige in *Wang Chun'e Teaches*. Her Stepson and Xue Dingshan in *At the Bend of the Feng River*.❷

拙译 调查发现平民出身的孩子，其外衫的前片是下垂的，就是所谓的茶衣。在京剧《三娘教子》中，薛倚哥角色的扮演者，包括《汾河湾》中薛丁山的扮演者，都身着茶衣。

原文 Chayi (Lit. "tea" clothes) are so named because of their color. Tea is a necessity of life and therefore has come to symbolize the ordinary. In the Song dynasty, there was an edict that ordinary people could only wear a beige color, therefore light brown has been associated with the working class for hundreds of years. Chayi, clothes for workers, are made of cotton rather than silk. The jacket opens in the center front and has a black collar band piped in white that ends in the fungus-shaped arabesque usually reserved for women's garments. The incongruity is considered a visual joke, as the characters who wear these garments are often comic. The sleeves of the jacket are three-quarter length and cuffed as though they have been rolled up for duty. The trousers are similarly shortened and cuffed. The cuffs on both the jacket and the trousers are white. The style of garment comes from lower-status wear that was modified for stage usage.

---

❶ PAN XIAFENG. The Stagecraft of Peking Opera ［M］. Beijing：New World Press，1995：125.

❷ 同❶120.

Chayi indicate poor men, often boatmen, woodsmen, or fishermen. In addition to browns and beiges, Chayi may also come in blue, such as the big sleeved servant robe (Daxiu). They are not embroidered.❶

拙译 茶衣因其颜色而得名。茶作为生活的必需品，因而成为普通人的象征。在宋代，法令规定普通人只能穿米黄色，因此数百年来浅褐色就与劳动阶级联系在了一起。茶衣作为劳动者的服装，材质为棉布而非丝绸。其样式为前开襟，黑色领口，配有白边，女款茶衣有着蘑菇形的蔓藤图纹装饰。京剧中茶衣的穿着者大多是滑稽逗乐的角色，通常借助服装的不协调带给观众视觉上的幽默。茶衣的袖长是普通外衣的四分之三，好像已经挽起来要当班的样子；裤子也同样剪短并卷起。外衫和裤子上的收口处都选用白边。茶衣的设计风格是对来自下层人服装的修改，以便更好地适用于舞台表演。

茶衣暗指穷人，通常是船夫、木匠或渔民。除了棕色和米黄色之外，茶衣也会选用蓝色，如不加刺绣的大袖。

（十二）大袖文化内涵及其协调翻译

大袖主要为店家、酒保、禁卒、骡夫所穿着，而且这种服饰是京剧中丑角的主要装束。其英文描述和作者拙译如下。

原文 A blue jacket with cascading sleeves and a large lapel is typical of the apparel of Chinese working people in former times, and it has been modified to dramatic effect on the stage often by adding a white kilt (Yaobao). In *Wu Song Kills His Sister-in-law*, Wu Dalang (brother of the title role), dressed in such a jacket as most of the bartenders do. Clearly marked by its long sleeves, the garment is popularly known as Daxiu, literally "big sleeves".❷

拙译 在过去，大翻领的蓝色外衫是中国劳动人民的典型服装，现在经常通过添加白色褡裢（腰包）来体现舞台的戏剧效果。在《武松杀嫂》中，武大郎（剧中主角人物的兄长）像多数的酒保那样穿着蓝色外衫，其明显特点是长袖，也就是众所周知的大袖，字面意思是"大的袖子"。

原文 The big-sleeved robe (Daxiu) is distinguished by its sleeves, which are wider and longer than usual and are finished with water sleeves. The neckline crosses over to the right in a diagonal line and is finished with a wide collar band, similar to the Xuezi.The big-sleeved robe is

❶ BONDS ALEXANDRA B. Beijing Opera Costumes: the Visual Communication of Character and Culture [M]. Honolulu: University of Hawaii Press, 2008: 165.

❷ PAN XIAFENG. The Stagecraft of Peking Opera [M]. Beijing: New World Press, 1995: 120.

worn by waiters, innkeepers, and other commoners. Although large sleeves and water sleeves may seem to indicate elegance, the large sleeves of the big-sleeved robe can be used more humbly, to wipe off.[1]

拙译 大袖长袍（大袖）因其袖子独具特色而得名，该袍服袖子比普通袍服的袖子更宽、更长，并以水袖收尾。样式为交领右衽，领口宽边，类似于褶子。大袖长袍通常由侍者、店主和其他平民穿着。虽然大袖子和水袖似乎用来表示优雅，但也可以更"谦卑"地被用来擦拭。

第二节
京剧二衣箱配饰文化内涵及其协调翻译

（一）僧背心、道背心文化内涵及其协调翻译

僧背心是佛教人物——老僧、和尚、尼姑穿在褶子外面的大背心。一般为绿黄格子面料，长度到小腿部位。道背心是道教人物——道士、道姑穿在褶子外面的大背心。一般为白蓝格子面料，长度到小腿。下面的英汉对照版本阐述了该服饰。

原文 僧背心样式与对襟大坎肩相同。由墨绿绸制，黄绸镶腰，穿于黑色侉（快）衣外，主要为戴发修行的头陀穿用。穿时，配以侉裤，绦子束胸，腰束大带。如《蜈蚣岭》中的武松、《四杰村》中的消月。[2]

译文 Sengbeixin (Monk's Vest) is similar to the loose-fitting robe worn by Buddhist monks. It is made of green satin with a style of sleeveless coat, which has a front parallel opening and yellow silk that be decorated decorations around the waist. It is worn by mendicant monks who submit to Buddhist disciplines. Silk ribbons are used to act like a kind of corset and belts are used to tighten the waist. Wu Song in *Wugong Mountain* (*Wugongling*) and Xiao Yue in *Si Jie Village* (*Sijiecun*) wear this kind of vest.[3]

---

[1] BONDS ALEXANDRA B. Beijing Opera Costumes: the Visual Communication of Character and Culture [M]. Honolulu: University of Hawaii Press, 2008: 166.

[2] 阙艳华, 董新颖, 王娜, 等. 中英文对照京剧服饰术语 [M]. 覃爱东, 陶西雷, 校译. 北京: 学苑出版社, 2017: 86.

[3] 同[2]86, 略有改动.

（二）绦子、大带文化内涵及其协调翻译

绦子是指用丝编织成的截面为圆或扁平的带子，可以系束或镶饰衣物。大带也被称作是鸾带，是夸张与美化后的腰带，全长6米，宽13厘米左右，由丝线或棉线织成，两头各有约33厘米长的丝穗。色彩多种多样，有花、素之分。穿箭衣、英雄衣时均需系大带。系法有丁字式、巾字式与介字式。下面的英汉对照版本更为详细地阐述了该服饰。

原文 "鸾带"又称"大带"，是一种由丝线或棉线编织而成的束腰软带，是为戏曲演员提供表演技巧、表达情绪的舞蹈工具。鸾带全长一丈八尺，宽三四寸，两端各有一尺长的排须状丝穗。演员穿抱衣抱裤、快衣快裤，或者箭衣的时候都系有鸾带。❶

译文 Luandai (Phoenix Belt) is also known as Dadai (Large Belt). It is made of silk or cotton, and is a soft belt that acts like a girdle. Intended to help the performer to express feelings, it is a dance appliance of highly aesthetic interpretation. It is six meters in length, and ten to thirteen centimeters in width, with whisker shaped silk tasselsat each end. Performers usually tie it while wearing Baoy and Baoka (tight-fitting jackets and trousers), Kuaiyi and Kuaiku (embroidered combat outfits) or Jianyi (archer's uniform). ❷

原文 丝绦俗称"绦子"，用丝线或棉线编织而成，全长约两丈，取中系八宝攒，结下有一条两尺半长的绦穗。使用时，先依男左女右的原则将绦穗定位在后腰部，再将左右绦绳顺时针绕至前腰正中部位，系蝴蝶结固定。丝绦主要为文人、谋士、师爷、塾师、秀才使用，亦作皇后、王妃的素服腰饰。❸

译文 Sitao (Silk Belt) is an accessory commonly known as Taozi (Ribbon). It is a rope knitted from silk or cotton, 6.67 meter in length, which has two silk fringes 83 centimeter in length at each end. It is fixed around the back of the waist according to the Chinese custom of "male wears it on the left, female on the right". Then the two fringes are tied around the front of the waist in a bow. Scholars, counselors, private advisers and private teachers have a silk belt; queens and princesses in plain clothes also have a silk belt for decoration. ❹

（三）侉子文化内涵及其协调翻译

原文 "侉子"由布料制成，样式呈长方形，下沿四周为扁叶状锯齿形，腰际束箍。

---

❶ 阙艳华，董新颖，王娜，等. 中英文对照京剧服饰术语［M］. 覃爱东，陶西雷，校译. 北京：学苑出版社，2017：88.
❷ 同❶88，略有改动.
❸ 同❶47.
❹ 同❶47，略有改动.

服色有黄、黑、白、绿等。黄色，上绘毛纹，表示猴皮，叫作"猴侉子"；黑色，上绘火焰，表示鬼火，叫作"鬼侉子"；白色，上绘波纹，表示水族；绿色，上绘叶脉，表示神话角色。侉子有时与云肩配套使用。❶

译文 Kuazi (Decorative Waistband) is an accessory in the shape of rectangle. It is made of cotton cloth. The waist is tied with a hoop and the clothing edges are zig-zag shaped. Different colours and designs are used to indicate different things. For example, a yellow Kuazi with the pattern of fir represents monkey skin and is called "Monkey Kuazi", A black Kuazi painted with flames symbolises fire, and is known as a "Ghost Kuazi". A white Kuazi painted with water ripples represents aquatic animals. Green Kuazi with patterns inspired by the veins of leaves represent mythological characters.❷

本章小结 ｜ 本节归纳总结了京剧二衣箱服装（猴甲、卒坎肩、大铠、茶衣、大袖）及其配饰（僧背心、道背心、绦子、大带、侉子）的文化内涵，对照不同版本的英汉表达后发现：在实际的翻译过程中，译文与原文之间都在不同程度地求真，国内外学者无论是用汉语或英文表述京剧服饰，都在最大程度地传递京剧服饰的文化信息；同时，中外学者在其实际的表述中，都在尊重原文的基础上进行了换位思考，充分体谅读者的接受程度和理解力；更为重要的是在实际的表述中，因为京剧服饰特殊的审美可以带给读者丰富的联想，以及服饰色彩、款式、质地等对于人物身份的阐释等特殊作用，其英汉表达都注意传递京剧服饰本身的美学意蕴。综上所述，京剧服饰的翻译需要在"真善美"的文化协调中，实现文化传递，增进相互理解，彼此吸纳，丰富发展。

---

❶ 阙艳华，董新颖，王娜，等. 中英文对照京剧服饰术语［M］. 覃爱东，陶西雷，校译. 北京：学苑出版社，2017：87.
❷ 同❶87，略有改动.

第五章

京剧服饰三衣类文化内涵及翻译协调美

　　三衣是京剧服装中的水衣、胖袄、护领、小袖、大袜、彩裤（以上属于软片类），以及厚底靴、薄底靴、登云靴、皂鞋、彩鞋、鱼鳞鞋、打鞋、小孩鞋、僧鞋、花盆底鞋和福字履（以上属于硬类）等物品的统称。这些物品或衬于"大衣""二衣"之内，保护和衬垫大衣和二衣。尽管其质料、纹饰和做工等远不及大衣和二衣考究，但在塑造人物形象与刻画人物性格、身份，以及赋予人物褒贬等方面同样发挥着重要作用。

第一节
软片类服饰文化内涵及其协调翻译

1. 水衣文化内涵及其协调翻译

　　水衣是贴身穿的衬衣，因为演出过程中（特别是武场）经常被演员的汗水浸透，故俗称水衣子。水衣多用白色或月白色棉布制作，贴身穿用以保护大衣、二衣、胖袄、绸缎刺绣的袍服等服装，具有洗涤方便与价格低廉的特点。按其款式，水衣又分为男式和女式两种。男式水衣为大襟、斜领，尺寸宽松、舒展，在前后襟的腋下处各缀有一根约17厘米长的布带，穿好后再将两根布带系在一起。女式水衣为对襟、小立领，贴身合体，搭接处有扣子或扣眼。其相关英文与汉译简述如下。

　　原文　One of the accessories available for nearly every performer on the stage is the sweatshirt, or Shuiyi. As a cloth-made undergarment, it helps keep the players from being drenched with perspiration.❶

　　拙译　京剧舞台上几乎每位演员都需要的配饰就是汗衫、水衣。汗衫用棉布制作，顾名思义，其穿用的目的在于防止演员在表演过程中汗流浃背。

　　原文　水衣，又称"汗衫"。斜领，小袖口，袖长齐腕，由白色棉布制成。穿时不分

---

❶ PAN XIAFENG. The Stagecraft of Peking Opera［M］. Beijing：New World Press，1995：128.

行当，贴身穿用以间隔戏衣，可用来防止戏服沾染汗渍。❶

译文 Shuiyi (Water Shirt) is a white shirt made of cotton cloth. The rough cloth can absorb water and sweat. Hence it be placed directly next to the body, it prevents expensive costumes from being damaged by perspiration.❷

2. 胖袄文化内涵及其协调翻译

胖袄选用两层白布或月白色布做成，中间絮有棉花，并用线缝好以防因为棉花错动而变形，一般穿在水衣与大衣、二衣之间，对大衣和二衣起到支撑和衬垫的作用，使其具有一定的"挺劲"。搭好后的胖袄一般为大领、无袖，设计的肩头宽度要比常人的肩宽大出许多。胖袄的前小襟、前大襟和后身各成一片，只在肩头处相连。在前大襟中腰处钉两根约66厘米长的小布带，以便搭好后将前襟和后身拦腰束紧。

胖袄属于舞台人物的外用服装，其款式设计和尺寸大小不仅要考虑穿用某种服装人物的身份和场合，还要考虑这类人物的体型，以期塑造出完美的艺术形象。根据其尺寸的大小和厚薄，特别是根据其肩头的形状，可以分为大、中、小、女四个不同种类。其相关的英文表达如下。

原文 大胖袄也叫折肩胖袄，因其保管时可由肩部对折起来而得名，肩头外缘接近方形，肩部宽度要比常人的肩宽出约2~3寸，肩头围长约一尺，以便搭好后自然垂到演员腋下的高度。前襟长约2尺，到演员腹部；后身长2尺3寸到2尺4寸之间，到演员的臀部。胖袄的厚度各处并不完全相同，肩头部位最厚，并用粗线纳实，胸部以下逐渐减薄，后身下摆附近又逐渐加厚至半寸左右，以增加人物的臀围。演员只有搭上大胖袄才能挑起宽大的蟒袍、开氅等服装。❸

拙译 The big fat jacket is also called the shoulder and shoulder fat "Ao", because it can be folded by the shoulders when it is stored.The outer edge of the shoulder is nearly square. The width of the shoulder is about 2~3 inches wider than the normal shoulder, and the circumference of the shoulder is about a foot long, so that it can reach the height of the actor's armpit naturally. The front is about 2 feet long and goes to the belly of the actor. The length of the back is about 2 feet, 3 inches to 2 feet and 4 inches, to the butt of the actor. The thickness of the jacket varies from place to place. The shoulder is the thickest area and thick thread is used to make it solid. It gradually thins below the feet and thickens to about half an inch around the hem of the back to increase the hip circumference. Only by putting on the big fat "Ao" can the actor hold up the big robe and cloak.

❶ 阙艳华，董新颖，王娜，等. 中英文对照京剧服饰术语［M］. 覃爱东，陶西雷，校译. 北京：学苑出版社，2017：91.
❷ 同❶91，略有改动.
❸ 孔祥芸. 京剧服装中的"三衣"（上）［J］. 戏曲艺术，1993(4)：85–89，略有改动.

原文 中胖袄又称圆肩胖袄，它与大胖袄的主要区别是肩头的形状不同，呈圆弧形。此外，其围长略小于大胖袄，厚度和肩宽与大胖袄相同；前襟和后身的长度略小于大胖袄。厚度从胸部以下逐渐减薄，后身下摆处不再加厚。搭中胖袄可塑造肩宽、腰细、像扇面似的体型，通过外穿箭衣、身背绦子、腰系大带会形成威武剽悍、浑厚有力的艺术形象。❶

拙译 Medium-fat "Ao" is also called round shoulder fat jacket. The main difference between it and big-fat jacket is the shape of shoulder, and the shoulder of medium-fat coat is circular arc. The girth is slightly smaller than the big fat coat, and the thickness and shoulder width are the same as those of big fat jacket. The length of the front and back is slightly smaller than that of the big fat jacket. The thickness gradually decreases from below the chest, and there is no thickening at the lower part of the rear body. Wearing a medium-fat "Ao" can make the shoulder width, waist thin, shape fan-like, and wearing an arrow coat, the body back, waistband will form a powerful and powerful image of art.

原文 小胖袄的样式与中胖袄基本相同，其肩头呈圆弧形，但肩部的宽度较小，比常人的肩宽只宽出1~1.5寸，厚度较薄，比较柔软、服帖。前襟和后身也略小于中胖袄。根据胖袄的厚度，小胖袄又分为厚小胖袄和薄小胖袄两种。厚小胖袄供长靠武生、文武老生在扎靠、穿箭衣、穿蟒时使用。薄小胖袄供小生、老生在穿褶子、帔、老斗衣，以及短打武生穿侉衣、抱衣时使用。❶

拙译 The style of the small fat "Ao" is basically the same as that of the medium fat "Ao". The shoulders are circular, but the width of the shoulders is small. The width of the small fat jacket is only 1-1.5 inches wider than that of ordinary people. The thickness is thin, soft and sticky. The front and back are also slightly smaller than the medium fat "Ao". According to the thickness of the fat "Ao", the little fat "Ao" is divided into two kinds: thick fat "Ao" and thin fat "Ao". The thick fat "Ao" is used for changkao Wu sheng, Wenwulaosheng, when wearing Jianyi (Arrow Clothing) and Mang. Thin little fat "Ao" for young men, middle-aged men when wearing, Xuezi, Pei, Laodouyi, or for the Duanda Wusheng who wears Kuayi and Baoyi.

女胖袄肩头也呈弧形，薄厚和小胖袄差不多，圆形领口且无领。除了肩头垫絮较厚，身上只有薄薄一层棉花，类似夹袄。前襟与后身也比小胖袄略短。因女演员本身臀部大，所以后身下摆处不需要再垫厚。

---

❶ 孔祥芸. 京剧服装中的"三衣"（上）［J］. 戏曲艺术，1993（4）：85–89，略有改动.

女胖袄的英汉表述如下：

原文 A cotton-padded waistcoat, usually made of white cloth, is suitable particular for the "painted face" character who, as a man of power or wealth, needs to display bulk by suing such a piece as an undergarment. The coat is rather heavily padded around the shoulders. Its Chinese name is Pang' ao. Stunt men and old men roles may also wear padded waistcoats, but they are not as bulky. ❶

拙译 絮棉背心通常由白布制作，特别适用于花脸人物，这种有权或有钱的人物，常常需要通过这样的贴身衣来展示其大块头。这件外套背心肩部有垫肩，在汉语里被称为胖袄。武生和老生也可以穿有衬垫的马甲，但没有那么大块的。

原文 The padded vest (Pang'ao, lit. "fat jacket") is a theatrical convention, and it balances the increased height that comes from the platform shoes and the elaborate headdresses. The amount of padding on the chest and shoulders varies with the nature of the role; Jing wear the largest and the broadest shoulders and Xiaosheng use the smallest. The vest is open on the sides and closes with a crossover flap in the front, with ties at the waist.❷

拙译 衬垫坎肩（又称胖袄）在戏剧中属于程式化的表演服装，它平衡了来自厚底鞋和精心设计的头饰所增加的高度。胸部和肩部的填充量因角色而变化，净角选用肩部最大和最宽的坎肩，而小生则选用最小的。坎肩的款式为两侧开口，前面是斜襟，腰部用带子系住。

3. 彩裤文化内涵及其协调翻译

戏曲演员在舞台上穿的各种颜色的绸缎裤子，统称彩裤。在戏曲服装中，彩裤处于从属地位，其颜色取决于大衣和二衣。彩裤的式样根据裤脚的形式可以分为散腿和系带两种，散腿为穿矮帮鞋履时所用，而系带彩裤为穿靴子时所用。彩裤尺寸肥大，以利于演员表演。彩裤的上端是用宽约26.6厘米的白布做的裤腰，腰围一般为13.3厘米多。裤腰以下是用质地轻柔光亮且不易起皱的各种面料制成。按照面料的颜色，彩裤可分为大红彩裤、青彩裤和杂色彩裤三大类；按照彩裤上是否绣有花卉与图案，又可分为素彩裤和绣花彩裤两大类。其英文表述如下。

原文 The style of unfitted trousers used for traditional Jingju is taken from the trousers that were worn in daily life, which had been absorbed from the steppe nomads into the Chinese

---

❶ PAN XIAFENG. The Stagecraft of Peking Opera [M]. Beijing: New World Press, 1995: 128.

❷ BONDS ALEXANDRA B. Beijing Opera Costumes: the Visual Communication of Character and Culture [M]. Honolulu: University of Hawaii Press, 2008: 201-202.

wardrobe around the third or fourth century BC. The trousers are made of two tubular legs, with some additional shaping in the crotch, but the hip and waist circumferences are about the same. Both the men's and women's versions are made essentially one size fits all. The trousers for stage use are made of silk in most cases, with a straight wide waistband of muslin or similar cotton fabric sewn to the top of the joined legs, following earlier construction techniques.

Plain inner trousers are called colored trousers (Caiku), while the embroidered trousers worn either inside or outside have many names, depending on the usage.

All traditional Jingju characters wear trousers of some kind, as did all Han people. The top of the trousers has a wide, open waistband that is folded to fit and secured with a string tie. Lower-status males wear trousers, also tied and tucked into their boots or socks, or rolled and cuffed to indicate hard work. Soldier's trousers are visible under their short jackets. Women's trousers are either concealed by their skirts or long gowns, or are visible when worn with shorter jackets. Their hems are left to hang straight.❶

拙译 传统京剧中肥大的裤子样式源于日常生活中的裤子，在公元前3或4世纪左右才进入中国人的衣柜，汲取草原游牧者服饰之精华。裤子有两条裤腿，胯部形状特殊，但臀部和腰围大致相同。男性和女性的板式基本一致。多数情况下，舞台上的长裤由丝绸制成，遵循早期的制作工艺，选用平直的棉布或类似棉布的面料制成宽腰带，并缝在两腿连接处的上部。

普通裤子被称为彩裤，而无论穿在里面或外面的绣花裤子都根据其使用场合有许多不同的名称。

所有传统的京剧人物和汉族人一样，都穿裤子。裤子上面还有宽大的、散开的腰带，可以折起来并用绳子系住。地位较低的男性穿裤子时，常把裤腿塞入靴子或袜子中，或卷起裤脚边以示努力工作。常见士兵上穿短上衣，下配裤子。女裤要么被裙子或长袍遮住，要么搭配短袖外套，裤线笔直。

原文 彩裤，绸质裤子，尺寸肥大，裆深，便于演员做繁复的表演动作。颜色以黑、红居多，也有白色、粉色、湖色、皎月色、古铜色等。❷

译文 Caiku (Coloured Trouser), made of satin, has a deep crotch and come in a large size, allowing the actors freedom of movement to undertake complicated dance routines. Trousers can be black, red, white, pink, lake blue, bright moon and bronze in colour, though red and black are most often used. ❸

---

❶ BONDS ALEXANDRA B. Beijing Opera Costumes: the Visual Communication of Character and Culture [M]. Honolulu: University of Hawaii Press, 2008: 192.

❷ 阙艳华, 董新颖, 王娜, 等. 中英文对照京剧服饰术语 [M]. 覃爱东, 陶西雷, 校译. 北京: 学苑出版社, 2017: 91.

❸ 同❷91, 略有改动.

4. 护领文化内涵及其协调翻译

护领是演员在演出中用的衬领，一般分为棉护领和毛巾单片护领两种。棉护领是用棉布缝制成的长约83厘米，宽10厘米的布袋，里面絮上一层棉花，用绗缝固定好，以免棉花错动。两端钉上约1米长的小带，以便护领交叉搭好后可拦腰系紧。使用时外面套上漂白布或白色的确良布做的布套，以便可以经常取下来换洗，保持洁净。而毛巾单片护领是将白素毛巾折成三层用来代替棉护领，另外做一个漂白布或白色的确良布的夹片（长度和宽度基本和棉护领一样），两端缝上约1米长的白小带。其英汉相关表达如下。

原文 The inner collar (Huling) is a rectangle of white cloth that is wrapped around the neck, crossed in front left over right, and tied around the waist. The inner collar has an additional string on the center back of the neck that hangs to the waist, where it is caught by the waist ties. During the Ming dynasty, the emperor Hongwu (1368−1398) ordered all ladies-in waiting to wear a paper collar guard. Under many of the costumes, a Jing actor wears a heavily padded vest and white collar.❶

拙译 内领（护领）是指用长方形白布交叉围在脖子上的一种衬领，领子前部是左压右，并围腰扎住。内领中有另外一条带子位于颈部后面中间位置，垂悬并在腰部打结。明朝时期，洪武皇帝（1368—1398）命令所有侍奉的宫女都必须佩戴纸护领。在许多戏服的下面，净角演员都穿着白领且有垫肩的坎肩。

原文 "护领"由白布缝制而成，可以在护领里面加薄棉垫以使颈部变粗。既能防止油渍沾污衣领，又可使演员胸廓及两肩显得匀称。护领讲究白净、骨立，护领片更要经常洗涤，以保持干净洁白，是演员着装要求的"三白"（护领白、水袖白、厚底靴底白）之一。❷

译文 Huling (Protective Attachments) is a kind of lining attachment made of cotton cloth. Cotton pads can be added into the interior of the Huling to make the neck appear thicker. Besides preventing grease stains from soiling the collar, it also can make an actor's thorax and shoulders look more balanced. Huling should always be pristine white, and so the cotton pads inside the Huling need to be cleaned frequently. It is an essential part of what is called the "Three Whites", that is, a white Huling, white Water Sleeves and Boots with thick white soles.❷

5. 袜子文化内涵及其协调翻译

袜子可分为大布袜和小布袜两种。大布袜俗称大袜，过去常用白布做成，现在都

---

❶ BONDS ALEXANDRA B. Beijing Opera Costumes: the Visual Communication of Character and Culture [M]. Honolulu: University of Hawaii Press, 2008: 201.

❷ 阙艳华，董新颖，王娜，等. 中英文对照京剧服饰术语 [M]. 覃爱东，陶西雷，校译. 北京: 学苑出版社, 2017: 94.

京剧服饰三衣类文化内涵及翻译协调美

第五章

改用白的确良府绸或漂白斜纹布来制作。为了结实，袜底用两层或三层布，并用缝纫机车缝成行，大布袜的袜勒高约40厘米，到膝盖处，袜勒顶端前头钉一根约1米长的白布小带。穿大布袜主要是为了避免脚底打滑，增加脚下的摩擦力。小布袜的袜底和袜勒都是单层布，袜勒很短，只到脚腕上约7厘米，袜勒顶端前头也有两根约67厘米长的白布小带。小布袜为旦角、老旦、彩旦穿用；但只要是穿彩鞋、彩旦鞋和福字履的演员，则必须都穿白色小布袜。其相关英汉表达如下。

原文 Socks (Baibu Dawa)

Some roles, including waiters and monks, wear a pair of socks made from a heavy white fabric. The socks are wide at the top, and the bottoms of the trousers legs are tucked into them. It is curved and trapezoidal to fit snugly around the lower leg; it reaches to mid-calf and fasten in back with Velcro.❶

拙译 袜子（大布袜）

京剧中的一些角色，包括侍者和僧侣，都穿用厚白布袜。这种袜子顶部宽松，呈弯曲状，贴合小腿，方便裤脚塞入，然后再在袜子的后部用尼龙搭扣固定。

第二节
硬类（靴鞋）等物品文化内涵及其协调翻译

京剧鞋靴与实际生活中人们所穿的鞋靴差异巨大，在一定程度上属于基于生活的工艺美术品，其运用不受朝代、四季、环境与气候条件的限制，但在最大程度上符合了人物的身份、造型与表演的需求。

（一）厚底靴文化内涵及其协调翻译

厚底靴也称为高勒厚底靴，靴勒至膝盖之下，为京剧中男性普遍穿用，上至皇帝下至底层人民。靴帮、靴勒多由缎料制成，靴底为用草纸压制的10~13.3厘米的厚底，且有一层耐磨的厚皮革，并在靴底四周涂成白色，前有坡头。厚底靴的特点是使剧中人物在增加高度的同时凸显，大方稳重、黑白分明的性格特点，亮丽但不笨重，与所

---

❶ BONDS ALEXANDRA B. Beijing Opera Costumes: the Visual Communication of Character and Culture [M]. Honolulu: University of Hawaii Press, 2008: 200.

穿的蟒、靠等服装协调。另外根据靴勒的颜色、材料、纹饰，厚底靴可分为青厚底靴、花厚底靴、猴厚底靴。青厚底靴为上述的黑色厚底靴；花厚底靴的样式也同青厚底靴一样，只是面料的颜色不同，并在鞋面上绣有花卉图案；猴厚底靴选用黄缎面料，上面绣有黑色猴毛团花纹，是专为孙悟空穿蟒时所用。其英汉表述如下。

原文 The platform boots or Houdixue are employed in many roles including those of people, old and young, civil and military, as well as "painted face" characters. Mostly made of cloth or stain, the boots reach to the knee and have flat white soles about two or three inches thick.

Similar to the platform boots are the so-called "tiger-head" boots or "Houtouxue", which are embroidered with tiger-head patterns on the insteps. The "tiger-head" boots are said to have been introduced to the stage first by entertainers from the south.

Another style of boots made of green stain with cloud patterns, called in Chinese Yuntouxue, is always worn by actors playing the part of Guan Yu. They resemble the platform boots both in outer shape and in the thickness of the sole.[1]

拙译 京剧中许多角色人物都会选用平底靴或厚底靴，包括年长和年少的、文职和武官，以及花脸人物。该靴主要由布料或缎面缝制，长及膝盖，白色鞋底、厚度约为5厘米或7.6厘米。

类似于平底靴样式的是所谓的虎头靴或猴头靴，在鞋脸前部绣有虎头图案。据说，虎头靴首先是由南方艺人引入舞台。

另一种绿缎制的靴子，配有祥云图案，汉语称其为云头靴，常为《三国演义》中关羽的扮演者穿用，在外形和鞋底厚度上都类似于厚底靴。

阙艳华等学者对于厚底靴也有着如下的汉英对照描述。

原文 厚底靴，又称"高方靴""高底靴"。长筒、齐头，底厚约二至四寸。由黑色缎面或平绒做帮，前脸和靴帮等处绣花纹图案。靴腰正中缝有线带，穿时在腿上系牢。与蟒、靠、官衣、开氅等夸张的戏曲服装相搭配，以衬托生、净角色庄重、威严的气概。厚底靴分男、女两种，女用厚底靴样式同男靴。靴底厚度和靴色均由穿用者根据具体情况自己设定，主要用于女扮男装的角色。[2]

译文 Houdixue (Thick-soled Boot), also known as Gaofangxue (Men's High Boots with Thick Soles), are one kind of knee-high boot. With black satin or flannel uppers and square toecaps, Houdixue are embroidered with stripes or cloud patterns. A flannel welt

---

[1] PAN XIAFENG. The Stagecraft of Peking Opera [M]. Beijing: New World Press, 1995: 130.
[2] 阙艳华, 董新颖, 王娜, 等. 中英文对照京剧服饰术语 [M]. 覃爱东, 陶西雷, 校译. 北京: 学苑出版社, 2017: 95, 略有改动.

runs along the perimeter of the outsole. Two bands, which can be tied to the performers' legs, are stitched to the boots. Painted white, the thick soles may be 2~4 cun (Chinese inch, equals 1/3 decimeter) thick. They belong to Guanxue (Official Boots) characterized by artistic exaggeration and embellishment, which match the equally exaggerated costumes such as Mang (Imperial Robes), Kao (Armour), Guanyi (Official Robes), Chang (Cloaks), thus suggesting the solemnity and dignity of the figures. Women's boots share the men's style, and are mostly worn by female characters who are disguising themselves as men. The color and soles' thickness are generally determined by the actor wearing them.❶

1. 官靴文化内涵及其协调翻译

官靴的靴筒子较长，用青缎子做成，靴底为6.6~13.3厘米厚，白色。在现实生活里没有这样厚底的靴鞋。但是在京剧舞台上经过艺术加工，使靴底加厚，其作用是为了增加角色的身高、加重形象、突出身份。与其有关的英文表述如下。

原文 Official boots: These have thick soles to make the actors taller, and match the robes characterized by a high degree of artistic exaggeration.❷

拙译 官靴的鞋底较厚，是为了增加演员身高，从而与夸张的服装相协调。

阙艳华等国内学者对于官尖靴有如下的英汉对照表述。

原文 官尖靴，黑缎素面。靴脸呈尖角状，尖角部位靴底上翻。硬胎对脸，正中皮革滚口，靴勒高至脚踝，底厚约半寸，主要为番邦人物穿用。❸

译文 Guanjianxue (Ankle Boots with Upturned Toecaps), one kind of boot with hard inner lining, are similar to cloth shoes in real life. Made of plain black satin, they are ankle boots with soles of five fen (Chinese unit of length, equals 0.00333 metre) in thickness. Their openings are bound with leather and the toecaps are turned upwards.❹

2. 虎头靴文化内涵及其协调翻译

一般的厚底靴为素黑色，而虎头靴则可以使用其他颜色，且靴底较薄，并在靴的前脸饰以虎头形象和表示虎须的丝穗，在靴面和靴勒上则饰以虎皮纹理。用英文可以表

❶ 阙艳华，董新颖，王娜，等. 中英文对照京剧服饰术语 [M]. 覃爱东，陶西雷，校译. 北京：学苑出版社，2017：95.
❷ 赵少华. 中国京剧服饰 [M]. 北京：五洲传播出版社，2004：95.
❸ 同❶97.
❹ 同❶97，略有改动.

述为：A kind of boots decorated with tiger heads, usually worn by generals and military officers.

阚艳华等国内学者对虎头靴也有如下表述。

原文 虎头靴由缎料制成，是经过美化、装饰的改良靴，有薄、厚之分。靴尖正中镶虎头吞口，靴面绣虎纹图案。使用时与人物穿戴相配合，有两种类型的虎头靴，厚底虎头靴和薄底虎头靴。前者如绿缎虎头靴，可以在《华容道》里关羽（三国时期的大将）扮演的角色中看到；后者则会在身穿改良靠的武生（男性武士）身上再现。❶

译文 Hutouxue (Tigerhead Boots) are beautified Gailiangxue (Modified Boots) with urnamental patterns. Decorated with tiger heads in the middle of the toecaps, made of coloured satin and embroidered with patterns of tiger stripes, they should match the characters costume. There are two types thick-soled and thin-soled. The former can be found on Guan Yu (a great general in the Three Kingdoms period) who wears thin-soled shoes in green satin. The latter is used by Wusheng (Martial Male-role) actors whilst wearing in Gailiangkao (Modified Armour) .❶

（二）薄底靴文化内涵及其协调翻译

薄底靴多用皮革制作，薄底、矮勒〔至踝骨〕，靴面有青缎、青绒、青冲服呢三种，左右两片靴帮从正中镶皮口处缝起，其形状略似"骆驼鞍儿棉鞋"，靴尖用小皮包头。除素黑色的薄靴称为青薄底靴外，也有其他颜色并绣花和回纹边的彩薄底靴，以及由黄缎或土黄绵绸制成并绣有黑色花纹图案的猴薄底靴。靴子颜色和人物所穿衣物色彩相搭配。穿薄底靴人物的特点是干净利落、行动轻快，故也称快靴。一般武戏的角色，像武生、武旦、武净都穿快靴。国内外学者对此有如下相关表述。

原文 Footwear for combatants made of black satin is not only thin-soled but also lower than ordinary boots, so that it would be more proper to call them bootees although Chinese actors often dubbed them Kuaixue, literally, a pair of "quick boots." Ren Tanghui, one of the heroes in *Crossroads*, appears in bootees, so do Crane-Boy (Hetong) and Deer-Boy (Lutong), disciples of Immortal Polar Star, in *Stealing the Magic Herb*. Some of the flowery thin-soled bootees for actresses are usually also satin-made but in rich red, white, green, light blue, and so forth.❷

---

❶ 阚艳华，董新颖，王娜，等．中英文对照京剧服饰术语［M］．覃爱东，陶西雷，校译．北京：学苑出版社，2017：97，略有改动．

❷ PAN XIAFENG. The Stagecraft of Peking Opera［M］. Beijing: New World Press，1995：131.

拙译 用黑缎制作的快靴比普通靴子薄，而且鞋靿低，虽然中国演员经常称之为"快鞋"，但实际更适合称为"靴子"，其字面意思应为"快靴"。如京剧《三岔口》中的英雄任堂惠，以及《盗仙草》中的鹤鹿两仙（鹤童和鹿童）都是脚穿快靴。女演员的快靴也属于薄底，虽然也用缎子制作，但相对花哨，颜色丰富，有红色、白色、绿色和浅蓝色等。

原文 "快靴"又称"薄底靴"。薄底，软胎，半高靿，靴靿齐踝。前脸，线码勾纹，正中滚皮口，靴尖精缝小皮包头。此靴有花、素两种。花快靴在靴靿处绣有花纹，使用时与戏衣颜色、花形相配合；素快靴由青缎、黑布、青绒制成，在舞台上使用率较高，主要为武生、武净穿用。❶

译文 Kuaixue (Ankle Bootsith Thin Soleskle) is also named Baodixue (Thin-soled Boots). They are so named because they are thin-soled and boots with soft inner lining. Made of leather and soled with a blind stitch, there are stripes and cloud patterns embroidered on the vamps. The boot openings and toecaps are bound with leather. There are two types: colored and plain. Made of colored satin and embroidered with flowers, the colored shoes must match the rest of the costume. Worn by Wusheng (Martial Male roles), Wudan (Martial Female roles) or Wujing (Martial Painted Famale Roles) when they wear Kuaiyi Kuaiku (Jackets with lines of Densely Arranged Chinese-style Buttons along The Underside of the Sleeves and The Center at The Front, with Black Trousers) or Kaoyi Kaoku (Tight-fitting Jackets and Trousers).❶

事实上，除去上述的薄底靴子，朝方又称方头靴，其本身也同样属于薄底靴子。

朝方没有厚底靴庄重，属于文丑行当专用的高靿靴，多用黑色缎料制成。靴尖部分略呈长方形。靴底厚约3.3厘米，四周涂白色，为丑行所扮演的官员、太监、文人等穿用。阙艳华等国内学者对此有如下相关表述。

原文 朝方靴，简称"朝方"。长筒、方头。由黑缎或素色棉布制成，底厚不足一寸。舞台上尚未出现厚底靴时，人物均穿朝方，不穿挂摆服装的方巾丑和行动比较轻巧的人物也常穿此靴。主要为官吏、太监等文丑角色专用。❷

译文 Chaofangxue (Square Toecap Boot) is made of plain black satin or cotton. Painted white, the sole of each boot is less than 1 cun (Chinese inch equals 1/3 centimeter) in thickness. They are mostly used earlier in the play before the actor appears wearing Houdixue

❶ 阙艳华，董新颖，王娜，等. 中英文对照京剧服饰术语 [M]. 覃爱东，陶西雷，校译. 北京：学苑出版社，2017：99，略有改动.

❷ 同❶96.

(Thick-soled Boots). Nowadays, however, they are mainly worn by low ranking officials or eunuchs played by actors of Wenchou (Civil Comic Role). Fangjinchou (Comic Role with Square Kerchief) characters not in robes, or characters who need to be brisk, also wear this kind of boot.❶

（三）登云履文化内涵及其协调翻译

登云履的基本样式与云头厚底靴相仿，但鞋帮高不过踝骨，鞋面前端缀有用彩色缎子或缎絮上缝有棉花制成的云头，鞋后跟也缀有云头状的彩缎饰物。其寓意为脚踏五彩祥云，超脱凡世。鞋底是用草纸压制成的，最底层是皮革。厚度为3.3~10厘米，厚薄不等。鞋面有果绿色、粉色、皎月色、湖色、紫色、香色、黄色等多种颜色，根据扮演人物的行当，按常规选用。登云履主要供得道成仙的神话人物使用。阙艳华等国内学者对此有如下相关表述。

原文　登云履又称"如意履""拳头抱""拳头鞋"。样式同福字履，底较厚，约二寸。前脸镶饰凸起的双云套头图案，鞋靿饰回云勾纹图案。主要为仙官、道家及有法术的人物穿用。❷

译文　Dengyunlü (Shoes Decorated with Cloud Patterns) are so named because two raised ornaments with cloud patterns are decorated on the vamps. They are similar to Fuzilü (Shoes Embroidered with Chinese Character "Fu", meaning blessing or happiness). Their soles are about two cun (Chinese inch, equals 1/3 decimetre) thick. The uppers are also decorated with the same cloud patterns. They are worn by immortals, Taoist priests or half-god characters.❷

（四）皂鞋文化内涵及其协调翻译

皂鞋也叫方口皂，样式与现实生活中的圆口青布鞋相仿。鞋底用草纸压成，厚为2~2.7厘米，在草纸底下边再缝上皮革底或麻线纳成的底子。鞋面用青缎、青绒、青冲服呢制作，青缎和青冲服呢用于丑角扮演的贫苦百姓、皂隶、衙役、解差等下层人物。阙艳华等国内学者对此有如下相关表述。

原文　"方口皂"，方头、长脸，底厚约八分。有青缎面、黑布面两种。主要为一般百

❶ 阙艳华，董新颖，王娜，等. 中英文对照京剧服饰术语［M］. 覃爱东，陶西雷，校译. 北京：学苑出版社，2017：96，略有改动.

❷ 同❶98，略有改动.

姓、差役、仆从等人物穿用。另一种厚底方口皂与方口皂样式相同，唯鞋底略高。❶

译文 Fangkouzao (Cloth Shoes for Common People or Yamen Runners) are made of plain black satin or cloth, with square toecaps, round openings and longer vamps. The soles are eight fen (Chinese unit of length, equals 2.66 centimetre) thick. They are worn by commoners, especially yamen runners or servants.❷

（五）彩鞋文化内涵及其协调翻译

彩鞋是京剧中的一种尖口鞋，鞋底用皮革或轮胎胶底制作，鞋面用各色彩缎、绉缎制作，并且绣有或拉有各种花卉和昆虫图案，鞋子前端缀有各种彩色丝穗。彩鞋是京剧中青衣、花旦、闺门旦等女性人物穿褶子、蟒、古装、裙袄和裤袄时配套穿用的一种鞋。彩薄底鞋则是女性武打演员（武旦）的通用之物。国内外学者对此有如下相关表述。

原文 The actor and troupe leader Wei Changsheng (1744–1802) introduced a technique for imitating the look and movement of bound feet, called Caiqiao (lit. "stepping on stilts"). A device was used to support the wearer on tiptoe, similar to the en pointe position in ballet, though that practice has generally fallen into disuse.❸

拙译 演员、戏班班主魏长生（1744—1802）按照缠足方法创造出束脚的外观和动作，称彩跷（意为踩高跷）。这是一种用来帮助穿戴者踮起脚尖的鞋子，类似于芭蕾舞鞋中的尖头位置，不过这种做法现在已经不再使用。

原文 彩鞋是一种女用普通便鞋，是旦角穿用之鞋的总称（除老旦、彩旦外）。长脸，薄底，鞋尖缀一绺彩色丝穗，后跟稍高。颜色有红、蓝、粉、湖等色。使用时与角色服装相配合，主要为夫人、小姐、丫鬟等角色穿用。❹

译文 Caixie (Color Shoes) are common, thin-soled shoes with longer vamps for females. Their uppers are a little higher. Made of satin and embroidered with designs of fowers, they are red, light blue, pink or lake-blue. A cluster of tassels with distinct colours are used to decorate the toecaps. The colours of shoes and their tassels must match the costume. Characters, such as dignified wives, well-bred young girls or maids all wear coloured shoes.

---

❶ 阚艳华，董新颖，王娜，等. 中英文对照京剧服饰术语 [M]. 覃爱东，陶西雷，校译. 北京：学苑出版社，2017：99.
❷ 同❶99，略有改动.
❸ BONDS ALEXANDRA B. Beijing Opera Costumes: the Visual Communication of Character and Culture [M]. Honolulu: University of Hawaii Press，2008：48.
❹ 同❶103.

"Caixie" is also used as an umbrella term for all shoes worn by female roles.[1]

（六）鱼鳞洒（洒鞋）文化内涵及其协调翻译

鱼鳞洒的鞋底厚度与厚底靴相同，一般用草纸和皮底压制，而鞋帮以青缎为面料。鞋子的外观好似两片鱼鳞合在一起。鞋子上面有白丝线拉绣的鱼头、鱼眼和鱼鳞。水路英雄一般都穿鱼鳞洒。阙艳华等学者对此表述如下。

原文 洒鞋，又称"鱼鳞洒鞋"。样式与生活中的普通洒鞋相同，薄底，矮勒，鞋脸较长，左右对称镶饰鱼眼，鞋面上绣鱼鳞纹或其他图案。鞋面为布面或彩色缎面，蓝白相间。主要为渔夫等角色使用，如《打渔杀家》中的萧恩。[2]

译文 Saxie (Cloth Shoes for Soldiers or Fishermen) are worn by fishermen, and are low-cut, thin-soled shoes with a longer vamp. They are made of satin or cloth, and decorated with a symmetrical pattern of a fish eye on the vamps. The uppers are also decorated with fish scale designs or other patterns. Furthermore, Saxie with fish scale designs are also called Yulinsaxie ("Fish Scale Cloth Shoes"), and are worn by characters such as Xiao En in *The fishermen's Revenge* (*Da Yu Sha Jia*).[3]

（七）打鞋文化内涵及其协调翻译

打鞋是京剧中专供开打用的鞋，一种通常以轮胎胶底或皮革为鞋底的圆白缎子或棉绸鞋，鞋后跟上缀有鞋鼻，以便穿好鞋之后系上鞋带，绕过脚腕在脚面上系个蝴蝶结，既漂亮又跟脚。多数打鞋都配有裹腿，裹腿一般由缎面、白布里和中间的衬布组成，缎面颜色和图案与打鞋上的基本相同。裹腿长约33.3厘米，宽约26.6厘米，上宽下窄，沿纵向两侧各钉一排气眼扣，以便用白线绳束紧在小腿上。根据鞋面上的图案，打鞋又分为鱼鳞洒打鞋、哪吒打鞋、蛟打鞋、六丁六甲打鞋、《雁荡山》打鞋、黑白道打鞋（以上六种打鞋都带裹腿）、猴打鞋（带猴袜子）、水族打鞋和云头打鞋等九种。黑白道打鞋的鞋面用宽不足3.3厘米的黑缎和白缎相间缝制而成，裹腿亦然。阙艳华等国内学者对此有如下相关描述。

原文 打鞋由布料制成，圆口，矮勒。鞋面饰黑白相间且对称的斜条色道，主要用于

---

❶ 阙艳华，董新颖，王娜，等. 中英文对照京剧服饰术语［M］. 覃爱东，陶西雷，校译. 北京：学苑出版社，2017：103，略有改动.

❷ 同❶101.

❸ 同❶101，略有改动.

武将、武士、兵卒、家丁等翻打扑跌的角色。若于鞋脸处镶虎头，鞋帮绣虎纹图案，则为马童所穿。❶

译文 Daxie (Acrobatic Shoes) is one kind of low-cut shoe with a round opening, similar in shape to Kuaixue (Ankle Boots with Thin Soles). The shoes are made of cloth and decorated with black and white stripes. The back section of the shoe upper is also adorned with coloured silk. Daxie are for acrobatic characters such as warriors, soldiers, private guards, etc. They also suit stable boys when decorated with tiger heads on the toecaps and tiger stripes on the uppers.❷

（八）小孩鞋文化内涵及其协调翻译

原文 "小孩鞋"是京剧舞台上专供小孩穿用的一种鞋。鞋底为皮革或轮胎底；鞋面为皎月色或蓝色的大缎与绉缎，并用白丝绒线绣上鱼嘴、眼、鼻、腮等花纹；后跟上缀有鱼尾状布片，很像鱼鳞洒打鞋。小孩鞋的使用不受男女或贫富的限制。❸

拙译 Children's shoes are specially designed for children to wear on the stage of Beijing Opera. The sole is leather or tire rubber, the upper is bright moon color or blue, the fabric is satin or crepe satin, and is embroidered with fish mouth, eyes, nose, cheek and other patterns sewn by white velvet thread. The heel is decorated with fishtail-like cloth, which much like fish scales sprinkled shoe. The use of children's shoes is not limited by men or women, rich or poor.

（九）僧鞋文化内涵及其协调翻译

僧鞋是京剧中专供和尚、尼姑和道姑穿用的一种尖口鞋。鞋底是用草纸压成的，草纸底下还钉有一层皮革。鞋面采用灰、黑、紫、黄等颜色的单色棉绸和黑绒或黑缎制作，鞋面上没有任何饰物，表示佛门弟子清心寡欲。鞋沿有1.67厘米宽的鞋口。僧鞋和僧衣的颜色通常一致。阙艳华等国内学者对此表述如下。

原文 "僧鞋"又称"和尚鞋"。双脸，中间一道弯沟上翻，底厚约二寸。鞋勒处彩绣云纹图案。鞋后缝绸条，系在脚踝。颜色有多种，使用时与人物衣色相配合。主要为僧道穿用。❹

❶ 阙艳华，董新颖，王娜，等. 中英文对照京剧服饰术语 [M]. 覃爱东，陶西雷，校译. 北京：学苑出版社，2017：100.
❷ 同❶100，略有改动.
❸ 孔祥芸. 京剧服装中的"三衣"（下）[J]. 戏曲艺术，1994(02)：91-94，101，略有改动.
❹ 同❶101.

译文 Sengxie (Monk's Shoes) are worn exclusively by Buddhist monks or Taoist priests. The soles are two cun (Chinese inch, equals 1/3 centimeter) thick. A visible hook-like seam is sewn across the middle of each vamp. The uppers are decorated with cloud patterns. Two silk bands are stitched to the back part of each upper and can be fastened to the performer's ankles. A number of different colours are used, but they should always match the rest of the costume.❶

（十）花盆底鞋文化内涵及其协调翻译

花盆底鞋又称旗鞋，是京剧中女真族或满族妇女着旗装和旗袍时穿用的一种鞋。鞋帮的面料、颜色、花纹、鞋穗都和彩鞋一样，不同之处在于旗鞋的鞋底下面有一木托，根据木托的形状又分为花盆底和元宝底两种。花盆底鞋为京剧中的角色所穿，穿花盆底鞋的人物走路时两肩微微摇晃，为了保持平衡，胳膊需慢慢地左右甩动，两脚平起平落，给人慢条斯理、端庄优美的感觉。国内学者对此有如下表述。

原文 The soles of these shoes are thick and shaped a little like a Chinese flower basin. The shoes are meant for women in traditional Mandarin costumes. Wearing flower basin shoes, the actress needs to walk in steady steps, keeping the shoulders moving and the arms swinging slightly. This is not only for balance, but also grace and style.❷

拙译 花盆底鞋的鞋底很厚，形状像中式花盆，为女性穿着传统的满族服装而设计。女演员穿花盆底鞋时，需步履稳重，保持肩部与手臂的轻微摆动，这样不仅可以保持身体平衡，而且显得风度优雅。演员穿这种笨重的花盆底鞋走路，都需要经过特殊训练。

原文 "旗鞋"又称"花盆鞋""马蹄鞋"。其样式同普通彩鞋，只是鞋底另加有木底，呈倒置的花盆形，厚约二寸。面亦与一般彩鞋同，绣勾花草等图案。主要用于着旗装、梳旗头的公主、妃子等角色。另有船形底的旗鞋为《四郎探母》中的萧太后所穿。❸

译文 Qixie (Manchurian Shoes), also known as Huapenxie (Flowerpot Shoes), are similarin style to Caixie (Colour Shoes), but there is an additional wooden sole attached to each of them. The wooden sole is about two cun (Chinese inch, equals 1/3 decimetre) thick and shaped like a Chinese flower basin. The instep is embroidered with the pattern of grass or flowers. Qixie are used by female characters in Manchurian costumes such as princesses and

❶ 阚艳华，董新颖，王娜，等. 中英文对照京剧服饰术语［M］. 覃爱东，陶西雷，校译. 北京：学苑出版社，2017：101.
❷ 赵少华. 中国京剧服饰［M］. 北京：五洲传播出版社，2004：95.
❸ 同❶102.

imperial concubines. Furthermore, some characters such as Xiao Taihou (queen mother Xiao) in *Yang Yanhui Visits His Mother* (*Si Lang Tan Mu*), usually wear Qixie with boat shaped soles.❶

（十一）福字履文化内涵及其协调翻译

福字履也称为夫子履，是戏曲舞台上所用的一种鞋子。选用缎料制作鞋面，款式接近圆口便鞋。鞋子的前脸正中有福字或云纹等图饰。鞋底厚度约为3.3厘米或略多，底部钉有一层皮革底。颜色多为黑色，也有其他颜色，与戏服颜色相配。舞台上的老旦，不管贫富都穿夫子履，但一般有钱人选择香色夫子履，而贫苦人则穿蓝色或紫色夫子履。国内学者对此有如下相关表述。

原文 The old women's roles have another special model of shoes whose tips and vamps are conventionally stitched with cloud patterns in black thread against either a purple or a brown ground. The most notable character wearing such Fuzilü may be exemplified by the historical figure Grandma She (Shetaijun) of the Song dynasty (960−1279).❷

拙译 福字履属于京剧中一款特制的鞋，为扮演老妇人角色者穿用。通常是在紫色或棕色的鞋帮上用黑线绣制祥云图案。宋代（960—1279）《杨家将》中的佘太君，就是穿福字履的典型代表。

原文 "福字履"又称"蝠字履"。薄底，矮靿。多由秋香素缎制面，前脸正中镶"福"字、古钱、"万"字或蝙蝠套云头图案。主要为老翁、老妇、商贾、店主等穿用。❸

译文 Fuzilü (Character "fu" Footwear) are so named because they are always embroidered with Chinese character "Fu" or "Wan" (good luck), symbolized by the image of bats (bianfu the animal in traditional Chinese culture carries the meaning of a blessing). Made of satin, similar in form to ordinary cloth shoes in real life, Fuzilü are low-cut shoes, though their soles are comparatively thick. The soles are about eight fen (Chinese unit of length, equals 2.67 centimetre) thick. They are worn by elderly people, businessmen and shop owners.❹

❶ 阙艳华，董新颖，王娜，等. 中英文对照京剧服饰术语 [M]. 覃爱东，陶西雷，校译. 北京：学苑出版社，2017：102，略有改动.

❷ PAN XIAFENG. The Stagecraft of Peking Opera [M]. Beijing：New World Press，1995：131.

❸ 同❶98.

❹ 同❷98，略有改动.

第三节
京剧头衣的文化内涵及其协调翻译

京剧头衣指在京剧表演中不同人物角色头上佩戴的物品，因其制作材料以及佩戴人物身份、地位、年龄或行当不同，可将其分为盔、冠、帽、巾四大类。本节聚焦阙艳华等国内学者编著的《中英文对照京剧服饰术语》和美国学者亚历山德拉撰写的 *Beijing Opera Costumes* 中有关头衣术语的跨文化表达，剖析其汉语拼音音译与关照文化差异的意译表达，品味文化协调理论在京剧服饰翻译中的具体运用。

原文 Official men of court and generals wear sculpted and firm headdresses, while scholars and lesser soldiers wear soft fabric headdresses. The fabric headdresses are often made using the same fabric and embroidery designs as the garments they are worn with, while the golden filigree helmets can be worn with a variety of garments, although an effort is made to coordinate the color of the pompoms[1] (Rongqiu) and the tassels with the hue of the garment. The headdresses can be grouped in many ways, but four loosely defined categories will be used here: ceremonial headdresses and crowns (Guan), helmets (Kui), hats (Mao), and fabric caps (Bu, also called Jin). The connection between historical headgear and traditional Jingju designs is apparent in the some of the ceremonial headdresses, hat, and soft caps, but the helmets are an area of dress where theatricality has overtaken historical references.[2]

拙译 官吏和将领头戴有雕饰且坚硬的头衣，而学者和小卒的头衣则是软布制作的。头衣面料通常与衣服面料一致，包括刺绣设计。只有金丝头盔例外，虽然也需要尽量使绒球和流苏的颜色与头盔的色调相协调，但是总体来说，金丝头盔确实可以与各种服装相配。头饰有很多组别，大致可分为四类：礼仪冕或冠、盔、帽和巾。头衣的设计成为历史在传统京剧中的体现，甚至头盔已经成为服装领域重要的历史参照。

（一）盔

京剧表演中，武将在行军打仗时用盔来保护头部。盔属于硬胎，一般都缀有绒球、珠子等装饰物。盔类头饰分为两大类：一类为男将所佩戴，包括帅盔、倒缨盔、扎巾盔、罐子盔、荷叶盔、夫子盔、狮子盔、虎头盔、中军盔、八面威、林冲盔、贼盔；

---

❶ "pompom" 一词属于拼写错误，原文为绒球，正确拼写为 "pompon"，下文中全部做了改正。
❷ BONDS ALEXANDRA B. Beijing Opera Costumes: the Visual Communication of Character and Culture [M]. Honolulu: University of Hawaii Press，2008：249.

另一类为女将所佩戴，其种类与男将所佩戴的相比种类较少，主要包括女帅盔、女倒缨盔、七星额子和蝴蝶盔等。

1. 帅盔

原文　"帅盔"形似覆钟，前扇有面牌、额子、龙形飞翅耳子，后扇有一对龙形盖翅，冠顶吞口处有红缨三叉戟头，使用时加装绣龙小后兜。颜色有金、银两种。主要为元帅一类的人物所戴，如《霸王别姬》中的韩信、《凤还巢》中的洪功等。另有"老旦帅盔"，是在老旦凤冠的基础上加戟头、后兜，为挂帅出征的佘太君所戴。❶

译文　Shuaikui (Commander's Helmet) is worn by commanders in chief and is used to protect the head in battle. It has a small cloak with embroidered dragons on the back of the helmet. The front section is large, and the back section is curved into a hollow circle. It can be classified by color into silver and gold types, and is worn by characters such as Han Xin in *The Kings Bids Farewell to his Favourite* (*Ba Wang Bie Ji*), and Hong Gong in *The Phoenix Returns to Its Nest* (*Feng Huan Chao*). In addition, the helmet for the old female role, only ever worn by the character She Tai Jun, is called the Laodan Shuaikui (Helmet for old female commanders in chief), which is decorated with a headdress (Houdou) and a long-handled halberd on top of the helmet.❷

2. 女帅盔

原文　"女帅盔"样式呈半圆形，前扇为贴金小额子，正中配饰牡丹花面牌，左右镶大凤、光珠，组成凤戏牡丹图案。冠顶吞口处装有戟叉，使用时加饰红缎绣凤后兜。主要为统兵的女元帅所戴，如《穆桂英挂帅》中的穆桂英等。❸

译文　Nüshuaikui (Helmet of a Female General) is a decorative hat with a golden headband, peony pattern to the middle, and a phoenix with peals to both sides. It is worn by female generals, including Mu Guiying in the play *Mu Guiying Takes Command* (*Mu Gui Ying Gua Shuai*).❹

译文剖析　帅盔的直译可以是"Commander's Helmet"。其"commander"一词在《牛津高阶英汉双解词典》中的解释是"a person who is charge of sth., especially an officer in charge of a particular group of soldiers or a military operation，负责人；（尤指）

---

❶ 阙艳华，董新颖，王娜，等. 中英文对照京剧服饰术语［M］. 覃爱东，陶西雷，校译. 北京: 学苑出版社，2017: 148.

❷ 同❶148，略有改动.

❸ 同❶149.

❹ 同❶149，略有改动.

司令官，指挥官"❶。其简明扼要地阐释了帅盔属于军中人物（指挥官）所戴。此外，"helmet"一词在《牛津高阶英汉双解词典》中的意思是"a type of hard hat that protects the head，头盔；防护帽"❷。译者用"helmet"来翻译盔字，将帽子的材质和功用与原文完美对应。同样的帅盔，但是女帅盔被直译为"Helmet of a Female General"。《牛津高阶英汉双解词典》中对"general"一词的解释为"an officer of very high rank in the army and the US Air Force; the officer with the highest rank in the Marines，将军；（陆军、海军陆战队或美国空军）上将"❸。同是帅盔，却因其性别差异被分别翻译为"commander"和"general"，其实笔者认为可以统一翻译为"Commander's Helmet for Male"和"Commander's Helmet for Female"，或者全部用"general"代替"commander"也未尝不可。上下文同类京剧服饰术语的协调统一会更方便西方读者理解接受，更何况穆桂英挂帅原本就是要凸显女性与男性不相上下，在战场上巾帼不让须眉。

同时，原文中对于帅盔和女帅盔的描述详细、丰满，但是译文则相对简洁。例如"前扇有面牌、额子、龙形飞翅耳子，后扇有一对龙形盖翅，冠顶吞口处有红缨三叉戟头，使用时加装绣龙小后兜"在译文"It has a small cloak with embroidered dragons on the back of the helmet. The front section is large, and the back section is curved into a hollow circle"中凝练为前部很大，后部是一个空心的圆形物体。原文中"在老旦凤冠的基础上加戟头、后兜"协调翻译为"which is decorated with a headdress (Houdou) and a long-handled halberd on top of the helmet"，而女帅盔中的"左右镶大凤、光珠""冠顶吞口处装有戟叉，使用时加饰红缎绣凤后兜"则全部省译。

3. 倒缨盔

原文 "倒缨盔"的样式前低后高，前扇为大额子，后扇为圆锥形，在后扇加荷叶片，吞口处加红色或黑色倒缨，使用时需加后兜。倒缨盔有金胎、银胎两种，金色"倒缨盔"主要为彪悍粗犷的大将所用，如《淮河营》中的李左车、《华容道》中的周仓。银色"倒缨盔"为"三国"戏中的马超专用，又称"马超盔"。❹

译文 Daoyingkui (Ma Chao's Helmet), worn by agile and fierce generals, and is made with pasted golden silk. It is decorated on the back with a headdress (Houdou) made of black embroidered satin. The red vertical tassels dangling on the back of the helmet characterizes the Daoyingkui, which is worn by Li Zuoche in *Huaihe Camp* (*Huaihe Ying*), and Zhou Cang in *Huangrong Road* (*Hua Rong Dao*). If the helmet is white in color, that is, it is trimmed

---

❶ 霍恩比. 牛津高阶英汉双解词典［M］. 王玉章，赵翠莲，邹晓玲，等译. 7版. 北京：商务印书馆，2010：391.

❷ 同❶954.

❸ 同❶847.

❹ 阙艳华，董新颖，王娜，等. 中英文对照京剧服饰术语［M］. 覃爱东，陶西雷，校译. 北京：学苑出版社，2017：144.

with white pompoms❶, white embroidered ribbons and headdress, it is exclusively worn by Ma Chao and called the "Ma Chao Kui". ❷

译文剖析 译文舍弃了原文中倒缨盔样式的具体描述，包括前扇和后扇，简洁再现原文主旨，方便西方读者理解接受。同时，在译文中补充了倒缨盔在戏曲中的穿用细节，使译文整体结构适合西方读者的阅读要求。

4. 女倒缨盔

原文 "女倒缨盔"的样式同男倒缨盔。蒙黑绒，双边加牙子，缀电镀帽钉。弯月形耳子，垂两条白缎绣勾草图案的飘带，如《花木兰》中的花木兰、《荀灌娘》中的荀灌娘等。❸

译文 Nüdaoyingkui (Female Red-tasseled Hat) is a decorative satin hat, embellished with rivets, and decorated with a red tassel on the top, with a piece of bag-like cloth to the back. It is worn by female soliders. The color and pattern of the hat should normally match the costume. Examples of this costume can be found in ZhouFeng Ying❹ from *Hua Mulan*, and Xun Guanniang in *The Story of Xun Guanniang*.❺

译文剖析 倒缨盔和女倒缨盔分别适用于京剧中不同性别的人物。倒缨盔结合戏曲人物的名字翻译成"Ma Chao's Helmet"，在下文中给予了解释，属于马超专用。虽如此，但是大部分读者并不能理解马超是谁。建议译者补加上对于马超这个人物的解释。事实上，文中提到的京剧中其他人物也在佩戴倒缨盔，同样也没有对这些人物给以具体阐释。因此，笔者建议针对不同的读者群体，采取不同的翻译策略。对于相对了解京剧的西方观众或读者，译为"Ma Chao's Helmet"应该完全可以接受，但是对于不太了解京剧文化的普通西方受众群体，直接翻译为"helmet"也是一种理性的选择。对于是否需要补充关于马超这一人物的阐释，则需要看场合，如果是书面表达空间允许，则可以通过补注或尾注等方式。具体翻译方法的选择需要依据实际的语境或时间、空间等条件左右权衡，才能实现文化沟通交流的目标。

女倒缨盔的翻译与倒缨盔的翻译有所不同，作者除了忠实于原文采取音译的方式，又将其阐释翻译为"Female Red-tasseled Hat"。在《牛津高阶英汉双解词典》中，"tassel"的解释为"a bunch of threads that are tied together at one end and hang from cushions, curtains, clothes, etc. as a decoration（靠垫、窗帘、衣服等）的流苏、穗、缨"❻，较为生

---

❶ "pompom"一词属于拼写错误，原文为绒球，正确拼写为"pompon"，下文中全部做了更正。

❷ 阙艳华，董新颖，王娜，等. 中英文对照京剧服饰术语［M］. 覃爱东，陶西雷，校译. 北京：学苑出版社，2017：144，略有改动.

❸ 同❷144.

❹ 经核，"Zhou Feng Ying"并非《花木兰》中的角色，是印刷错误，应改为"Hua Mulan"。

❺ 同❷，略有改动.

❻ 霍恩比. 牛津高阶英汉双解词典［M］. 王玉章，赵翠莲，邹晓玲，等译. 7版. 北京：商务印书馆，2010：2069.

动地再现了"红色的流苏"这一特点。但是汉语中盔的本意是指用来保护头的帽子，多用金属制成，而"hat"在《牛津高阶英汉双解词典》中的解释为："a covering made to fit the head, often with a flat edge that sticks out, and worn out of doors"[1]，常指带檐的帽子。由此可见，译文与原文之间的意思并不对等，"hat"与"helmet"在材质方面差异明显。因此，笔者建议将其翻译为"Female Red-tasseled Helmet"。

5. 扎巾盔

原文 "扎巾盔"又称"硬扎巾"。由大额子前扇和圆形后扇组成。后扇突显硬火焰，形似扎巾。全盔为硬胎，铁纱漆底，有金、银两种。盔背装双龙立翅，下装如意形硬片后口。盔色与面部化妆色彩须和谐，如《挑华车》中的高宠、牛皋，《辕门斩子》中的孟良、焦赞，《战太平》中的陈英豹、王渊等。[2]

译文 Zhajinkui (literally "Towel Helmet") is one kind of helmet with a hard inner lining worn by generals. The back section resembles a towel, hence the name. It can be gold or silver in color, and the color needs to harmonize with the make-up of the actor. It is worn by roles such as Gao Chong and Niu Gao in *Tiao Hua Che*, Meng Liang and Jiao Zan in *Killing a Son in Yuanmen*, and Wang Yuan and Chen Yingbao in *War in Taiping*.[3]

6. 荷叶盔

原文 "荷叶盔"前扇扁圆，正中饰大绒球面牌，左右为龙尾耳子，缀有光珠、绒球。后扇上端似两瓣荷叶相对，盔背插双龙立翅。分金胎、银胎两种：金色荷叶盔缀红色绒球，银色荷叶盔缀蓝或紫色绒球。主要为武将所戴，如《战太平》中的花云、《失街亭》中的王平、《穆柯寨》中的杨延昭等。此盔于两耳子下加挂流苏，主要为太监专用，如《法门寺》中的刘瑾、《逍遥津》中的穆顺、《打龙袍》中的陈琳等。[4]

译文 Heyekui (Lotus Leave Helmet) is a headgear worn by military generals. The front section is flat and oval in shape, the middle is decorated with a large ball, and the tail is trimmed with beads and pompons. The back section resembles a lotus-leaf, hence the name. The helmet must match with the color of the rest of the costume and make-up. For instance, red pompons correspond with the gold helmet, while a silver helmet matches blue or purple pompons. This helmet is worn by such roles as Hua Yun in *War in Taiping*, Wang Ping in

---

[1] 霍恩比. 牛津高阶英汉双解词典［M］. 王玉章，赵翠莲，邹晓玲，等译. 7版. 北京：商务印书馆，2010：935.
[2] 阙艳华，董新颖，王娜，等. 中英文对照京剧服饰术语［M］. 覃爱东，陶西雷，校译. 北京：学苑出版社，2017：140.
[3] 同[2]140，略有改动.
[4] 同[2]141.

Losing the Battle in Jieting, and Yang Yanzhao in *Muke Villiage* (*Mu Ke Zhai*). If tassels dangle from the ears, this helmet can also be worn by favored imperial eunuchs, such as Liu Jin in *Famen Temple* (*Fa Men Si*), Mu Shun In *Xiao Yao Jin*, and Chen Lin in *Beating the emperor's Robe* (*Da Long Pao*).❶

7. 罐子盔

原文　"罐子盔"样式同帅盔，但较简化。盔顶装立叉且垂缨，盔背加三块瓦式后兜，贴金点绸。❷

译文　Guanzikui (Helmet in the Shape of Jar), similar to helmet for commander in chief, this helmet is worn by palace guards. It looks like a jar pasted over with gold-inflected silk, hence the name. The top of the helmet is decorated with a vertical fork, and tassels dangle from both sides.❸

8. 蝴蝶盔

原文　"蝴蝶盔"全盔贴银点绸，冠前装红色大绒球，冠顶镶大蝴蝶，左右有蝴蝶形耳子，挂桃红线穗。冠背有湖色光缎制，并绣飞蝶图案的后兜，使用时可插双翎、挂狐尾。主要为女将所戴，如《樊江关》中的薛金莲。❹

译文　Hudiekui (Butterfly Helmet) is a decorative silver helmet with pink pompons and a large butterfly on the top, with small butterflies and a pink tassel on each side. It is worn by female generals, such as Xue Jinlian in the play *Fanjiang Pass* (*Fan Jiang Guan*).❹

9. 狮子盔

原文　"狮子盔"全盔金胎点绢，前扇为小额子，后扇顶部为狮子头，使用时加挂黑缎绣狮子图样后兜。主要为爱说大话但有武艺的大将所戴，如《神亭岭》中的太史慈、《逍遥津》中的华歆等。❺

译文　Shizikui (Lion Helmet) is a kind of headgear with a decorated back section. The front section is smaller than the back section, which makes the helmet resemble the shape of

---

❶ 阙艳华，董新颖，王娜，等. 中英文对照京剧服饰术语［M］. 覃爱东，陶西雷，校译. 北京：学苑出版社，2017：141.

❷ 同❶143.

❸ 同❶143，略有改动.

❹ 同❶152，略有改动.

❺ 同❶154.

a lion. Decorated with the patterns of lions, the back of the helmet has a hood made of black satin attached to it. The helmet is usually worn by Taishi Ci in *Shenting Ridge* and Hua Xin in *Xiao Yao Jin*.❶

10. 虎头盔

原文 "虎头盔"前扇为小额子，后扇塑虎头形。虎面凸出正中的"王"字，双耳耸立，盔身与后兜饰虎毛纹路。戴此盔者，表示勇武善战，如《战宛城》中的典韦、《八大锤》中的何元庆等。若摘掉面牌，改插一根独翎，则表示战将的甲胄残缺，如《甘露寺》中的贾化。❷

译文 Hutoukui (Tigerhead Helmet) is worn by valiant generals and is decorated with patterns of tiger fur. The black section resembles a tiger head, within which the Chinese character "Wang" (literally "King") protrudes. It is used by roles such as Dian Wei in *War in Wancheng* (*Zhan Wan Cheng*) and He Yuanqing in *Taking Turns Fight against Four Paris of Mallets* (*Ba Da Chui*). If the Mianpai (a circular ornament used in costumes for acrobatic fighting roles) is replaced with one single pheasant tail (rather than two), it means the generals' amour is incomplete. This variant is worn roles such as Jia Hua in *Ganlu Temple* (*Gan Lu Si*).❸

译文剖析 鉴于以上五种头饰均以其形状命名，故笔者对其归纳剖析。扎巾在京剧头饰中分为两类，即硬扎巾和软扎巾。扎巾盔译为"Towel Helmet"，应该为硬扎巾。因为"helmet"与汉语中的盔相对应。但是，汉语巾多为纺织品，质地柔软，其本意是指包裹或覆盖东西的用品。查阅《牛津高阶英汉双解词典》"towel"一词的解释为"a piece of cloth or paper used for drying things, especially your body，毛巾；手巾；抹布；纸巾"❹。将其用在头上，确实可以指代包裹完好。扎巾盔形似扎巾，译为"Towel Helmet"，方便读者理解其形态。同样，荷叶盔、罐子盔、蝴蝶盔、狮子盔和虎头盔的翻译也都采用了相同的翻译策略，基于中文本意，在充分考虑其头饰的具体形制下，译者采用了直译的翻译策略，方便读者理解其形状、材质与角色人物身份等。

11. 八面威

原文 "八面威"又称"霸盔"或"八角冠"。由"倒缨盔"增饰而成，但尺寸较宽

❶ 阚艳华，董新颖，王娜，等. 中英文对照京剧服饰术语［M］. 覃爱东，陶西雷，校译. 北京：学苑出版社，2017：154，略有改动.

❷ 同❶155.

❸ 同❶155，略有改动.

❹ 霍恩比. 牛津高阶英汉双解词典［M］. 王玉章，赵翠莲，邹晓玲，等译. 7版. 北京：商务印书馆，2010：2140.

大。由前后两扇组成，前扇为大额子，正中缀面牌，左右对称镶饰金龙、雄狮，组成二龙戏珠或狮子滚绣球图案；后扇形似覆钟，外套八角形宽边。每角挂红缨一束。使用时加饰后兜，主要为称威称霸的人物所戴，如《霸王别姬》中的项羽、《将相和》中的廉颇等。❶

译文 Bamianwei (Tyrant's Helmet) is made of gold-infused silk and worn by domineering characters. Derived from the Daoyingkui (literally "inverted tassel helmet"), it is decorated on both sides with the patterns of two dragons chasing a pearl, or lions playing with a ball. The back section is shaped like an upside down clock, and is decorated with red tassels. It is worn by roles such as Xiang Yu in *The King Bids Farewell to His Favourite* (*Ba Wang Bie Ji*) and Lian Po in *The General and The Premier Reconciled* (*Jiang Xiang He*).❷

译文剖析 "八面威"意译为 "Tyrant's Helmet"。"tyrant"一词在《牛津高阶英汉双解词典》中的意思为："a person who has complete power in a country and uses it in a cruel and unfair way，暴君；专制君主；暴虐的统治者"❸。而 "八面威"的喻义为八面威风，即为称王称霸之人所戴，与 "tyrant"一词所指代的意思并不完全吻合。所以如果在特定语境下，能够结合其形制给予更多阐释性的翻译，则可能会加深读者的理解。

12. 林冲盔

原文 "林冲盔"又称 "夜奔盔"。由 "倒缨盔"改制而成，前口正中装牛心倒缨。全盔为黑胎，蒙黑色光缎或黑平绒，镶蓝色双边，缀电镀泡钉。使用时需加装后兜，主要为《逼上梁山》中的林冲专用。❹

译文 Linchongkui (Lin Chong's Helmet) is exclusively worn by Lin Chong in *Driven to Join the Liangshan Rebel*s (*Bi Shang Liang Shan*). It is made of black satin and velveteen, and has inlaid blue layers to both sides. This helmet is also worn by Su Wu in *Su Wu Herds the Sheep*.❺

译文剖析 林冲盔和马超盔都是基于京剧人物的名字来命名的，属于专人佩戴的头饰。但是译文与此同时还补充了《苏武牧羊》中的苏武也同样适用该头饰，容易使西方读者产生不必要的困惑。因此，笔者认为如此解释似乎多余，不妨将其省译。

---

❶ 阙艳华，董新颖，王娜，等. 中英文对照京剧服饰术语 [M]. 覃爱东，陶西雷，校译. 北京：学苑出版社，2017：154.

❷ 同❶154，略有改动.

❸ 霍恩比. 牛津高阶英汉双解词典 [M]. 王玉章，赵翠莲，邹晓玲，等译. 7版. 北京：商务印书馆，2010：2182.

❹ 同❶147.

❺ 同❶147，略有改动.

13. 贼盔

原文　"贼盔"前扇弯月形，侧视如桃状，桃嘴向前。盔前有火焰顶，镶异色生丝或白色兔毛，使用时需加装后兜。主要为江湖侠客、盗贼所戴，如《猎虎记》中的邹渊、邹润，《大破铜网阵》中的欧阳春。❶

译文　Zeikui (Thief 's Helmets) is worn by swordsman, such as Zou Yuan and Zou Run in *The Story of Hunting Tiger* (*Lie Hu Ji*), and Ouyang Chun in *The Tactic of Breaking the Net* (*Da Po Tong Wang Zhen*). Viewed from the side, the helmet resembles the shape of a peach, the back of which is decorated with rabbit fur. The front sections shaped like a curved moon, whilst the back section is a hollow and rounded.❶

译文剖析　汉语中的贼专指偷东西的人，而英语中的"thief"在《牛津高阶英汉双解词典》中的意思为："a person who steals sth. from another person or place，贼；小偷；窃贼"❷。从字面意思来看，译文"Thief 's Helmets"比较准确地表达了原文内涵，而且也方便译语读者理解其人物身份。但是贼盔在京剧中实际并不仅局限于盗贼佩戴，江湖侠客也常选用该帽饰，而贼和江湖侠客的中文含义差异巨大。因此如果直接将京剧中盗贼和江湖侠客两类截然不同的人物所佩戴的头盔都称为贼盔，必然混淆译文读者对于京剧人物身份的理解与认识。因此笔者建议将译文补充为"worn by both swordsman and thief"。

14. 中军盔

原文　"中军盔"形似礼帽，顶部装有锥形吞口，头盔下口有三寸宽檐。全盔金胎点绢，没有光珠和绒球，主要为中军所戴。❸

译文　Zhongjunkui (Main Forces Helmets) is one kind of helmet, and similar to the formal hat. It is worn by a major general in the military. The helmet has a three-inche wide brim with patterns of arrows made of three halberd leaves. It is gold in color and does not have and decorative pompons attached.❹

译文剖析　京剧中的中军是指军队的主力大部队，而"forces"一词在《牛津高阶英汉双解词典》中的意思为："the weapons and soldiers that an army, etc. has, considered as things that may be used，武装力量"❺。将"main forces"作为"helmet"的定语，可以

---

❶ 阙艳华，董新颖，王娜，等. 中英文对照京剧服饰术语 [M]. 覃爱东，陶西雷，校译. 北京：学苑出版社，2017：158，略有改动.

❷ 霍恩比. 牛津高阶英汉双解词典 [M]. 王玉章，赵翠莲，邹晓玲，等译. 7版. 北京：商务印书馆，2010：2097.

❸ 同❶159.

❹ 同❶159，略有改动.

❺ 同❸794.

方便译语读者了解该头饰与特定人物之间的关系。

15. 夫子盔

原文 "夫子盔"样式前扇低后扇高,上面小下面大。正中为二龙戏珠图案,左右龙耳子挂大穗、飘带。顶部大火焰的两侧有龙须,冠前有黑、蓝、白等色的绒球。白夫子盔为《长坂坡》中的赵云、《满江红》中的岳飞所戴,绿夫子盔为关羽所戴,黑夫子盔为项羽所戴。❶

译文 Fuzikui (Warrior's Helmets) is worn by warriors. The upper part is larger than the lower part, and the front part is higher than the back part. The helmet looks like an upside-down clock, and is decorated with tassels and ribbons on both sides in black, blue and white. The big red pompon on top of the helmet is trimmed with dragons chasing a pearl. The white warriors' helmet is worn by Zhao Yun in *Changban Slope* (*Chang Ban Po*) and Yue Fei in *The River all Red* (*Man Jiang Hong*). The green warriors' helmet is decorated with yellow pompons and a headdress (Houdou), and is worn by Guan Yu. The black warrior's helmet is for Xiang Yu.❷

译文剖析 在汉语中,夫子作为称谓表达,其含义可分为五种,一是古时对男子的尊称;二是旧时称呼学者或老师;三是旧时称自己的丈夫;四是称读古书而思想陈腐的人(含讥讽意);五是孔门的学生对孔子的称呼。其中第一种意思最符合原文中对于夫子的解释,即古时对男子的尊称。查阅《牛津高阶英汉双解词典》,"warrior"的意思是"a person who fight in a battle or war,(尤指旧时的)武士,勇士,斗士"❸。因此该译文体现了对于冠者的尊重与仰慕,凸显了剧中正面人物的形象。

16. 七星额子

原文 "七星额子"样式呈半圆形,冠前上端排两层大绒球,每层七个,每排大绒球下都插有光珠,中间有一排小绒球,左右凤耳子挂双排粉色珠穗。主要为女战将、女武官、女中军等身份的人物所戴,使用时可插双翎。如《红鬃烈马》"银空山"一场中的代战公主、《樊江关》中的樊梨花等。❹

译文 Qixing'ezi (Seven Starred Headband) is a decorative silver headband with two

❶ 阙艳华,董新颖,王娜,等. 中英文对照京剧服饰术语 [M]. 覃爱东,陶西雷,校译. 北京:学苑出版社,2017:142.
❷ 同❶142,略有改动.
❸ 霍恩比. 牛津高阶英汉双解词典 [M]. 王玉章,赵翠莲,邹晓玲,等译. 7版. 北京:商务印书馆,2010:2265.
❹ 同❶150.

158

layers of pompons, seven pompons on each layer. It is decorated with two rows of pearls tassels, and is worn by female soldiers, female officers and female generals, etc. Characters such as Princess Daizhan in the play *The Steed with the Red Mane* (*Hong Zong Lie Ma*) and Fan Lihua in *Fanjiang Pass* (*Fan Jiang Guan*) wear this headband.[1]

译文剖析 汉语中七星的本意是指头饰上的七个绒球，直译为"seven starred"是表面忠实于原文，但实际上并未阐释其文化内涵，且容易使读者误解为缀有七颗星星。原文中的额子是指无顶的头巾，一般为妇女扎头所用。在《牛津高阶英汉双解词典》中，"headband"的解释为："a strip of cloth worn around the head, especially to keep hair or sweat out your eyes when playing sports，头带，束发带（尤指运动时用以固定头发或吸汗）"[2]。因此，"seven starred"与"headband"的组合，在一定程度上可以较好地帮助西方读者理解其形制。虽然在中英文的解释上并非完全对等，存在一定的语意流失，但是站在术语翻译的角度，该译文保留了基本含义，且较好地传递了原文的内涵。

综上所述的十六种盔类，在使用时多数通常"需加装后兜"。经请教北京市非物质文化遗产项目"京剧盔头"的第五代传人李鑫老师和刺绣技艺大师安宁，后兜一般随着京剧剧装一起制作、刺绣，需与戏衣的颜色、纹样相匹配。软胎的盔头一般与后兜缝合固定，而硬胎的盔头则可依据服装的不同配以相应的后兜。或许正是基于该原因，译文对后兜的阐释给予了适度省略，以免读者混淆。查阅《新牛津英汉双解大词典》，"flap"的含义为"a piece of something thin, such as cloth, paper or mental hinged or attached on one side only, that covers an opening or hangs down from something，（一端固定的布条、纸或金属件等）片状垂悬物；片状封口物"[3]。所以如需对后兜进行补充翻译，笔者认为"flap"在一定程度上较为贴切，不妨尝试将其译为"a flap on the back of helmets"，再根据语境对后兜的色彩纹样进行具体阐释。

（二）冠

冠是指比较郑重的礼帽，为帝王和贵族所佩戴。本节将京剧中的冠依照佩戴者的身份依次分为九龙冠、紫金冠、翠凤冠、珠凤冠、如意冠、大过梁、老旦凤冠、小过梁、麻冠和道姑冠，并将其汉英表达对照剖析如下。

---

[1] 阙艳华，董新颖，王娜，等. 中英文对照京剧服饰术语［M］. 覃爱东，陶西雷，校译. 北京：学苑出版社，2017：150，略有改动.

[2] 霍恩比. 牛津高阶英汉双解词典［M］. 王玉章，赵翠莲，邹晓玲，等译. 7版. 北京：商务印书馆，2010：943.

[3] 牛津大学出版社. 新牛津英汉双解大词典［M］. 上海外语教育出版社，编译. 2版. 上海：上海外语教育出版社，2013：823.

1. 九龙冠

原文 "九龙冠"的周围共饰有九条龙，故称"九龙冠"。有软、硬之分。样式前低后高，前圆后平。颜色为金色，帽前正中缀一只大黄绒球，旁边点缀数十颗珠子。左右装龙形耳子，下垂黄色流苏。❶

译文 Jiulongguan (Nine-dragon Crown) is one kind of helmet, hat and cap. The surface of the crown is decorated with nine dragons, and is worn by emperors. It is made of embroidered yellow satin or gold-inflected silk. It has hard inner linings, with the front sitting lower than the back. It is decorated with a yellow ball and dozens of beads in the middle of the anterior part of the crown. The back of the crown bears two wings. Yellow tassels dangle from the left and right ears. ❷

译文剖析 龙在中国古代是帝王的象征，其地位神圣不可侵犯，正如我国古代帝王有"真龙天子"之称。九龙又来自"龙生九子"之说，作为饰物表示祥瑞。九龙直译为"nine-dragon"，这与西方邪恶象征的龙寓意相悖。因此，在译文的解释中，如果文化场合的时空允许，则不妨补加有关中国龙文化内涵的阐释，以方便西方读者准确把握该帽饰的象征意义。同时，查阅《牛津高阶英汉双解字典》，"crown"的解释为："an object in the shape of a circle, usually made of gold and precious stones, that a king, or queen wears on his or her head on official occasions, 王冠；皇冠；冕"❸。由此可见，英汉语言中的"crown"与皇冠含义基本对等，但是九龙的文化内涵在英汉语言中差异明显。当然我们也有理由相信，随着对外交流的不断扩大，西方邪恶与怪兽喻义的龙与汉语中尊贵天子龙的寓意将逐渐得到理解与接受，从而使术语的翻译更加灵动雅洁，充满美感。

2. 紫金冠

原文 "紫金冠"又称"太子盔"或"太子帽"。前扇为额子，后扇原形头盔顶上加"垛子头"（都子头）。盔上有绒球、珠子若干，两侧龙耳尾子挂长流苏。盔背下端挂如意形垂穗硬片，名"后口"（后遮根）。紫金冠分银胎和金胎两种，主要为皇太子、世子王孙、功勋后裔、少年将领所戴。戴金色紫金冠者，如《孙安动本》中的徐龙、《四平山》中的李元霸；戴银色紫金冠者，如《穆柯寨》中的杨宗保、《八大锤》中的陆文龙。❹

译文 Zijinguan (Purple Gold Crown) is worn by princes, princesses, and the

❶ 阙艳华，董新颖，王娜，等. 中英文对照京剧服饰术语 [M]. 覃爱东，陶西雷，校译. 北京：学苑出版社，2017：130.
❷ 同❶130，略有改动.
❸ 霍恩比. 牛津高阶英汉双解词典 [M]. 王玉章，赵翠莲，邹晓玲，等译. 7版. 北京：商务印书馆，2010：480.
❹ 同❶164.

descendants of military families. The front section sits forward, and the helmet on the back section is decorated and trimmed with decorative balls and dozens of beads, from which tassels dangle. Most crowns have silver silk, hanging pink pompons and tassels; but some have gold silk, red pompons and tassels. The golden purple gold crown is worn by roles such as Xu Long in *Sun An is Angry* (*Sun An Dong Ben*) and Li Yuanba in *Siping Mountain* (*Si Ping Shan*). The silver purple gold crown is worn by Yang Zongbao in *Muke Village* (*Mu Ke Zhai*), and Lu Wenlong in *Taking Turns Fight against Four Paris of Mallets* (*Ba Da Chui*).❶

译文剖析　紫金属于合金，与玫瑰金类似，是彩金的一种。其主要成分是黄金、铜、镍和钴等金属，可用来制作首饰，是尊贵、财富或地位的象征。紫金直译为"purple gold"，方便西方读者理解；"crown"一词则表明了佩戴者的身份高贵，简洁准确地传递了原文信息。

3. 翠凤冠

原文　"翠凤冠"样式如折扇的扇面，以玲珑的点翠立凤为主要装饰，故称"凤冠"。冠前饰米珠，上镶九只小凤，冠顶有五只大凤，凤嘴衔串珠，中间的珠串短，两边的珠串长。凤冠两侧挂串珠排挑，凤尾耳子下垂挂黄色珠穗，冠后加饰硬片彩穗后兜。主要为正宫皇后或具有皇后身份的人物所戴，如《逍遥津》中的伏后、《大宝国》中的李艳妃等。❷

译文　CuiFengguan (Phoenix Coronet) is a decorative fan-shaped coronet with a little pearl and nine small phoenixes on the front, and five larger phoenix on the top. It is worn by an empress or women of high rank. Examples of this costumes can be found in the character of Fu Hou from *Xiao Yao Jin*, and Li Yanfei in *Defending the Nation* (*Da Bao Guo*).❶

译文剖析　在我国传统文化中，龙、凤相似，凤属于中国古代传说中的百鸟之王，象征祥瑞。古代帝王嫔妃常佩戴带有凤的饰物，象征其地位之高贵。在《牛津高阶英汉双解词典》中，"phoenix"的意思是："a magic bird that lives for several hundred years before burning itself and then being born again from its ashes,（传说）中的凤凰，长生鸟"❸。同样，查阅《牛津高阶英汉双解词典》发现："coronet"的释义为："a small crown on formal occasions by princes, princesses, lords, etc.（王子、公主、贵族等戴的）冠冕"❹。"coronet"与翠凤冠中冠的含义相近，但是译者并没有将点翠这一中国传统金

❶ 阙艳华，董新颖，王娜，等. 中英文对照京剧服饰术语［M］. 覃爱东，陶西雷，校译. 北京：学苑出版社，2017：164，略有改动.

❷ 同❶164.

❸ 霍恩比. 牛津高阶英汉双解词典［M］. 王玉章，赵翠莲，邹晓玲，等译. 7版. 北京：商务印书馆，2010：1487.

❹ 同❸445.

银首饰的特殊制作工艺给予传递，而是采取节译，导致其在原文的忠实程度上还有待于在未来跨文化的交际中逐步完善。

4. 珠凤冠

原文 "珠凤冠"的样式同翠凤冠。冠前并排有五层蝴蝶与光珠，每层各有蝴蝶七只、光珠七对，两侧挂串珠排挑。左右为蝴蝶耳子，挂红色或粉色双排珠穗。冠后加装硬片彩穗后兜，光彩耀目。主要为皇后、贵妃、公主、郡主等人物所戴，如《铡美案》中的皇姑、《贵妃醉酒》中的杨玉环、《甘露寺》中的孙尚香等。❶

译文 Zhufengguan (Pearl Phoenix Coronet), similar to the Phoenix Coronet, but is decorated with five layers of butterflies and pearls, with seven butterflies and seven pairs of pearl on each layer. It is worn by empresses, princesses, concubines or women of high rank. Characters such as Huanggu in the play *Execution of Chen Shimei* (*Zha Mei An*), Yang Yuhuan in *The Drunken Beauty* (*Gui Fei Zui Jiu*) and Sun Shangxiang in tha play *Ganlu Temple* (*Gan Lu Si*) wear this coronet.❶

译文剖析 原文中的珠凤冠和翠凤冠很相似，但是冠上的装饰物有所不同。珠凤冠是以蝴蝶和珍珠为主要装饰，且都为七对；而翠凤冠则是具有点翠制作工艺的凤冠。译文中的 "Pearl Phoenix Coronet" 较好地对应了原文的珠、凤、冠三个汉字，从字面翻译评判，字字对应，简洁凝练，而从文化内涵的角度评判，则略显单薄。但是作为京剧术语，其译文需要充分考虑文化语境和场合，以及时空的局限与读者或观众的需求，再决定有无必要补充宝珠和凤的文化内涵。同时，随着国际交流的日趋频繁和中华文化对外传播的速度与变化，西方人对于珠和凤，包括点翠工艺的了解与接受也在不断加深，译者如何协调翻译也处于动态变化之中。

5. 老旦凤冠

原文 "老旦凤冠"样式同翠凤冠，但形制略小。冠前缀七只凤头，凤嘴衔串珠。冠顶镶大凤，两侧镶小凤，挂串珠排挑。主要为皇太后、诰命夫人所戴，如《铡美案》中的国太、《杨门女将》"金殿"一场中的佘太君、《打龙袍》中的李宸妃等。❷

译文 Laodanfengguan (Phoenix Coronet for Elderly Women) is a decorative coronet of golden color, similar to the Phoenix Coronet, but with a larger phoenix on the top, a small phoenix on each side. It is also adorned with seven phoenix heads. Each phoenix holds a stream of pearls in their mouth. It is worn by the Empress Dowager and ladies with honorary

---

❶ 阚艳华，董新颖，王娜，等. 中英文对照京剧服饰术语［M］. 覃爱东，陶西雷，校译. 北京：学苑出版社，2017：166.
❷ 同❶167.

title conferred by imperial mandate. Characters such as Guo Tai in the play *Execution of Chen Shimei* (*Zha Mei An*), She Taijun in *The Yang Family Generals* (*Yang Men Nü Jiang*), Li Chenfei in *Beating the Imperial Robe* (*Da Long Pao*) wear this coronet.❶

译文剖析 尽管老旦凤冠和翠凤冠的形制不同，但是都属于凤冠的一种，在翻译策略的选择方面颇为相似，两者都被直译为"Phoenix Coronet"。事实上，凤冠仅为皇后或具有皇后身份的人所佩戴，而老旦凤冠则为皇太后和诰命夫人所佩戴，补译为"Phoenix Coronet for Elderly Women"。查阅《牛津高阶英汉双解词典》，"elderly"的意思是"used as a polite word for 'old'，年纪较大的，上了年纪的"❷。因此，"elderly women"意为年长的女人。但在京剧中，老旦凤冠为皇太后或诰命夫人所戴，并非是普通年长女性可以随便使用的头饰，如果在原译文中再补充"high social position"之类的定语修饰"elderly women"，则在一定程度上能更彰显中国文化内涵。

6. 如意冠

原文 "如意冠"又称"虞姬冠"，因形似如意而得名。样式前低后高，冠底装有托口，托口上装三层如意片子。如意片前端呈如意云头状，后端微翘，周围垂挂串珠。左右饰黄缎飘带，上绣云纹或嵌光片。主要为梳古装头的王妃、太后一类的人物所戴，如《淮河营》中的吕雉、《霸王别姬》中的虞姬等。❸

译文 Ruyiguan (Ruyi Crown) is a decorative yellow satin crown, with a wand in the shape of an "S" usually made of jade (which symbolizes good fortune). The crown is embellished with streams of pearls and yellow satin ribbons. It is worn by princesses, empresses dowagers or concubines. Examples can be found in the characters of Lü Zhi in the play *Huaihe Camp* (*Huai He Ying*), and Yuji in the play *The King Bids Farewell to His Favorite* (*Ba Wang Bie Ji*).❶

译文剖析 如意原意是指呈S形的器物，材质多为玉，象征吉祥如意。由于如意属于中华特色词汇，在英语中没有对应的器物，只能将其名称音译为"Ruyi"，再结合具体内涵给予阐释，帮助读者进一步理解。

7. 大过梁

原文 "大过梁"亦称"大过桥""大过翘"或"半凤冠"。样式呈扁圆形，饰有点翠

---

❶ 阚艳华，董新颖，王娜，等. 中英文对照京剧服饰术语［M］. 覃爱东，陶西雷，校译. 北京：学苑出版社，2017：167，略有改动.

❷ 霍恩比. 牛津高阶英汉双解词典［M］. 赵翠莲，译. 8版. 北京：商务印书馆，2014：644.

❸ 同❶167.

立凤或珠凤，上缀光珠与小绒球，左右凤耳子挂排子穗或花篮穗，使用时加饰如意形硬片彩穗后兜。主要为剧中王妃、公主或偏妃之宫冠，如《逍遥津》中的曹妃。❶

译文 Daguoliang (Large Bridge) is a kind of helmet, hat or cap, shaped like an oval. It is worn by concubines or princesses, such as Concubine Cao in *Xiao Yao Jin*. Covered with gold silk, it is decorated with jade phoenix, small pompons, and small phoenix dangling from either side of the temples.❷

8. 小过梁

原文 "小过梁"亦称"小过桥"或"小过翘"。样式呈扁圆形，上缀光珠与小绒球，左右凤耳子挂流苏，后无兜。四顶为一堂，用时卡在大头上。主要为剧中宫女的头饰。❸

译文 Xiaoguoliang (Small Bridge) is a kind of helmet, hat or cap. It is in the shape of a hollow oval and is worn on the head by maids. It is decorated with bright small balls and small pompons, and small phoenix on either side from which tassels dangle.❹

译文剖析 梁在汉语中的本意是泛指水平方向的长条形承重构件，也指器物、身体或其他物体上中间高起的部分。译文7和8中的大过梁和小过梁是指中间高起配以珠宝的头饰，而译文再次将其协调为"helmet"（头盔），虽然似乎冠（crown）更突显其装饰效果，但实际又都与原文有一定的差距，容易误导西方读者。或许随着时间的推移，在未来直接音译为"Liang"更符合其本意。

9. 麻冠

原文 "麻冠"箍式。形状如小型额子，不装面牌，不加耳子。贴银点绢，缀黑色小绒球。使用时头戴水发，额上裹一道白色绸条，系结于脑后。此冠在戏中表示披麻戴孝，如《战冀州》中的马超、《战樊城》中的伍员、《九江口》中的张定边等。❺

译文 Maguan, decorative crown with silver lines and pearls worn by performers in deep mourning, is usually accomplished the dress in white and with burlap draped over one's shoulders in mourning for parent. Such examples can be seen as Ma Chao in the play *Battle at Jizhou* City (*Zhan Jizhou*), Wu Yuan in the play *Battle at Fancheng City* (*Zhan Fancheng*), Zhang Dingbian in the play *JiuJiang Estuary* (*Jiujiang Kou*) .❺

---

❶ 阚艳华，董新颖，王娜，等. 中英文对照京剧服饰术语［M］. 覃爱东，陶西雷，校译. 北京：学苑出版社，2017：178.
❷ 同❶178，略有改动.
❸ 同❶179.
❹ 同❶179，略有改动.
❺ 同❶174.

译文剖析 我国自古就有"披麻戴孝"之说，意为长辈去世，子孙需身披麻布服，头上戴白，表示哀悼。麻冠是指在京剧中人们参加丧礼时所佩戴的头饰。麻作为服饰材质，中外皆有，英文的对应词汇为"linen（麻布，亚麻布）"。虽然将麻冠直接英译为"Maguan"方便记忆中文语音，但是忽略了西方读者本可以通过该术语对于其材质或形制理解的机会。在跨文化交际过程中，相信随着读者对于中华京剧服饰文化理解程度的加深，类似译文也将逐渐被理解接受。

10. 道姑冠

原文 "道姑冠"，样式上窄下宽，前口中凹，两侧弯云边，佩戴时卡在大头上。冠背挂十八节云层式垂穗后兜，颜色有杏黄、皎月等。主要为年轻的女道姑及女僧所戴，如《双下山》中的色空、《秋江》中的陈妙常等。❶

译文 Daoguguan (Cap of a Taoist Nun) is a decorative black cap with gold edging and embroidered with peonies. To the narrow section at the front, a curved pattern symbolizing clouds is embroidered. The cap can be either apricot yellow or bright moon in color. It is worn by young Taoist nuns and Buddhist nuns. Characters such as Se Kong in the play *A Nun Craves Worldly Vanities* (*Shuang Xia Shan*), and Chen Miaochang in *Autumn River* (*Qiu Jiang*) wear this cap.❷

译文剖析 译者将道姑冠直接音译为"Daoguguan"的同时，也提供了"Cap of a Taoist Nun"这样的阐释性译本。冠的本意为帽子，此处协调翻译为"cap"。"nun"在《牛津高阶英汉双解词典》中的解释为"a member of a religious community of women who promise to serve God all their lives and often live together in a convent, 修女"❸。原文中的道姑属于中国道教中的女道士，类似于西方修女。因此，"Taoist nun"在表明其是中国道教头饰的同时，也巧用"nun"表明了佩戴者的身份。虽然在中国传统文化中，道姑冠不仅为道姑所佩戴，还为尼姑所佩戴，而尼姑属于中国佛教。但"道姑冠"的译文并没有涉及尼姑，在一定程度上并不完全符合佩戴者身份，还有待于基于语境作深度协调。

（三）巾类

巾属于一种家常便帽，种类繁多，分为软胎和硬胎。根据《中英文对照京剧服饰术语》中关于巾的术语翻译，本节尝试将其分组剖析如下。

❶ 阙艳华，董新颖，王娜，等. 中英文对照京剧服饰术语［M］. 覃爱东，陶西雷，校译. 北京：学苑出版社，2017：168.
❷ 同❶168，略有改动.
❸ 霍恩比. 牛津高阶英汉双解词典［M］. 王玉章，赵翠莲，邹晓玲，等译. 7版. 北京：商务印书馆，2010：1368.

1. 皇巾

原文 "皇巾"又称"帝王巾"。软胎，样式呈方形，前低后高，帽背插朝天翅一对。帽身由黄缎制成，彩绣草龙、回纹等图案。主要为皇帝睡卧或染病时所戴。❶

译文 Huangjin (Emperors "Cap") is also known as Diwangjin ("Diwang" means emperor). It has a soft inner lining, a front section lower than the back section, and a pair of wings in the middle of the back. It is made of yellow satin and is embroidered with dragons. It is worn by an emperor when he is in the bed or becoming ill.❷

译文剖析 皇在汉语中的本意为君主，或者是神话传说中的神。皇巾中的皇意思是皇帝或皇上，即中国古代最高的统治者。巾字的原意是指擦东西或包裹、覆盖东西的用品，多为纺织品。皇巾质地属于软胎，由最能代表皇室的黄缎制成，彩绣草龙象征此配饰仅由皇帝或皇上佩戴。因此，皇巾音译为"Huang Jin"与阐释性的译文"Emperors 'Cap'"。"emperor"在《牛津高阶英汉双解词典》中的解释为"the ruler of an empire，皇帝"❸；而"cap"在该词典中的解释为"a soft hat that fits closely and is worn for a particular purpose，软帽"❹。由此可见，"cap"的解释和巾的中文意义在翻译上虽并不对等，但是在京剧中人们常用巾来指代便帽，而且皇巾又指皇帝在睡卧或生病的特殊场合才佩戴的头衣。在一定程度上，该译文有助于京剧服饰的文化传播。

2. 相巾

原文 "相巾"由缎料制成，样式呈方形，前脸正中镶白玉帽正一块，帽背插朝天翅一对，高出帽顶。帽身彩绣龙、蟒及垂直纹。颜色有紫、红、蓝等。主要为达官显贵及宰相穿便服时所戴，如《宇宙锋》中的赵高。❺

译文 Xiangjin (Premier's Kerchief) is a kind of square cap with several raised lines and a hard inner lining. It is made of satin and embroidered with dragons, pythons or patterns of golden lines. It has a pair of wings attached to the back and a piece of white jade that decorates the middle of the front. It is worn by prime ministers, the nobility and high officials when they are in everyday dress, such as Zhao Gao in The Sword (*Yu Zhou Feng*).❻

译文剖析 相巾在京剧中专指达官显贵及宰相穿便服时佩戴的头饰。在《古代汉

---

❶ 阙艳华，董新颖，王娜，等. 中英文对照京剧服饰术语［M］. 覃爱东，陶西雷，校译. 北京：学苑出版社，2017：107.

❷ 同❶107，略有改动.

❸ 霍恩比. 牛津高阶英汉双解词典［M］. 王玉章，赵翠莲，邹晓玲，等译. 7版. 北京：商务印书馆，2010：653.

❹ 同❸283.

❺ 同❶108.

❻ 同❶108，略有改动.

语词典》中，相原专指辅佐君主的大臣，后专指宰相❶。相巾协调翻译成"Premier's Kerchief"。"premier"在《牛津高阶英汉双解词典》中的解释为："used especially in newspapers, etc.to mean 'prime minister',（尤用于报纸等）首相，总理"❷。"kerchief"的意思为："a square piece of cloth worn on the head or around the neck，方头巾；方围巾"❸。巾字含义相对宽泛，协调翻译为"kerchief"可在形制方面一定程度上方便西方读者理解接受。但是，作者将相翻译为"premier"，容易误导西方读者把相巾理解为只为宰相佩戴。

3. 员外巾

原文 "员外巾"由缎料制成，样式呈斜方形，帽背有两根软翅，下垂绣花飘带。帽身彩绣草龙或"寿"字等图案，颜色有紫、蓝、古铜色等。主要为员外、乡绅、富户等人所戴。❹

译文 Yuanwaijin (Squire Cap) is a kind of ladder-shaped square cap with a soft inner lining. It is made of satin and is generally embroidered with the Chinese character "shou" (longevity). There are two soft wings and two ribbons embroidered with flowers attached to the back. The costume and Yuanwaijin worn by a character must be of the same color and embroidered pattern. Yuanwaijin is ususlly worn by rich men.❺

译文剖析 在《古代汉语词典》中，员外的意思是"正员以外的官员。后世可用钱捐买，故常用以称呼有钱有势的豪绅"❻。在《牛津高阶汉英双解词典》中，"squire"的解释是："(in the past in England) a man of high social status who owned most of the land in a particular country area,（旧时英格兰的）乡绅，大地主"❼，这与京剧中的员外并不完全对等。为了避免西方读者对其理解偏差，不妨在"squire"前面补加"Chinese"，方便其准确理解该头饰仅为中国乡绅所佩戴。

4. 文生巾

原文 "文生巾"又称"小生巾"或"公子巾"。由缎料制成，帽盔呈半圆形，自帽顶至两侧有扇形硬边，与左右如意头硬边相连接。帽背垂两条飘带，彩绣草龙或"万"字

❶《古代汉语词典》编写组. 古代汉语词典［M］. 北京：商务印书馆，2007：1707.
❷ 霍恩比. 牛津高阶英汉双解词典［M］. 王玉章，赵翠莲，邹晓玲，等译. 7版. 北京：商务印书馆，2010：1559.
❸ 同❷1109.
❹ 阙艳华，董新颖，王娜，等. 中英文对照京剧服饰术语［M］. 覃爱东，陶西雷，校译. 北京：学苑出版社，2017：109.
❺ 同❹109，略有改动.
❻ 同❶1944.
❼ 同❷1955.

（卍）盘边等图案，使用时与人物配色相配合。主要为儒雅的公子，清秀、潇洒的秀才和书生所戴。若两根飘带于左侧打结，则表示遇难或正在赶路。❶

译文 Wenshengjin (Civilian Kerchief), also known as Gongzijin (young urbane man's cap) or Xiaosheng jin (young man's cap), is one kind of cap with a soft inner lining. It is made of satin in various colors and embroidered with flowers, grass or the Chinese character "wan" (literally "myriad of things", a symbol of gook luck). It has a hard fan-like decoration on either side from which tassels dangle. Two ribbons are attached to the back. Sometimes the ribbons can be tied in front of the left ear in order to show the character is suffering misfortune or hurrying on his way. There are many colors of Wenshengjin and they must match the rest of the costume. Wenshengjin is usually worn by young men of refinement or handsome young scholars.❶

5. 武生巾

原文 "武生巾"样式同文生巾，唯帽顶正中插小火焰（俗名"软火焰"）且不加飘带，开打场时可摘去流苏，使用时与人物服色相配合。主要为武生（含武小生）所扮演的角色所佩戴。❷

译文 Wushengjin (Military Kerchief), one kind of cap with a soft inner lining, is similar to the "civilian kerchief ". The significant difference between the two is that the Wushengjin does not have ribbons attached. Instead, a "small flame" (a popular name for a bow made of red silk) is added to the center of the front. In acrobatic fighting performances, the tassels on either side must be removed. It is worn by Wusheng (the male role skilled in martial arts) characters and Wuxiaosheng (young male role who is also skilled in martial arts).❷

译文剖析 京剧中文生巾和武生巾彼此对立存在，但二者均为生这一角色使用。在《古代汉语词典》中，文生是对读书人的称呼❸。文生巾多为京剧中儒雅公子、秀才或书生等人物选用。根据《牛津高阶英汉双解词典》中的解释，"civilian"的意思是："a person who is not a member of the armed forces or the police，平民；老百姓；庶民"❹。由此可见，文生与"civilian"在意思上并不完全对等。而与此相对的武生，则指京剧中擅长武艺的角色。查阅《牛津高阶英汉双解词典》发现，"military"的意思是"connected with soldiers or the armed forces，军事的；军队的；武装的"❺，而且英文中并

❶ 阙艳华，董新颖，王娜，等. 中英文对照京剧服饰术语［M］. 覃爱东，陶西雷，校译. 北京：学苑出版社，2017：110.
❷ 同❶111.
❸《古代汉语词典》编写组. 古代汉语词典［M］. 北京：商务印书馆，2007：1393.
❹ 霍恩比. 牛津高阶英汉双解词典［M］. 王玉章，赵翠莲，邹晓玲，等译. 7版. 北京：商务印书馆，2010：345.
❺ 同❹1273.

没有与武生相对应的译文，直译为"military kerchief"容易使读者误认为该帽与军事相关。因此，武生与文生的译文在文化翻译层面都造成了流失与损耗，现阶段在一定程度上容易导致读者的误解。

6. 学士巾

原文 "学士巾"又称"解元巾"。由缎料制成，样式为前低后高，帽顶左右两端各有挖角，帽背插一对如意形软翅，分花、素两种，使用时与人物服色相配合。主要为有知识、有计谋但无功名的文人所戴。❶

译文 Xueshijin (Scholar's Kerchief), also known as Jieyuanjin (Jieyuan), is a scholar in ancient China who won the first place in provincial imperial examinations). It is made of satin and embroidered with flowers, with the color and pattern used determined by the rest of the costume. The cap has a tilt, so that the front part sits lower than the back part. It has two Ruyi-shaped (Buyi, an S-shaped ornament object symbolizing good luck) soft wings attached to the back. It is worn by renowned scholars.❷

译文剖析 查阅《古代汉语词典》，学士是知识分子的统称，与有无功名无关❸；在《牛津高阶英汉双解词典》中，"scholar"的意思是"a person who knows a lot about a particular subject because they have studied it in detail，学者"❹。而在中华传统京剧中，学士巾主要为有知识、有计谋，但无功名的文人所戴。因此，单从字面意思推断，"scholar"和学士并不完全对等，"Scholar's Kerchief"在一定程度上未能协调源语文化与译语之间的文化差异。对于原句中"如意"的英文表达则非常全面，从读者到形状与文化内涵，都在最大限度上满足了西方读者的阅读需求。

7. 老人巾

原文 "老人巾"样式同学士巾，前脸正中镶玉帽正，巾背有后兜和飘带，帽身彩绣云朵等图案，颜色有紫、黑、古铜三种。主要用于彬彬有礼的长者或慈祥的神仙。❺

译文 Laorenjin (Elderly Kerchief) is also named because it is worn by the elderly. It is made of satin with a soft inner lining and is without embroidery. It has a hood at the back and a jade "cap adjuster" to the front. Purple, black and bronze are the three colors mostly

---

❶ 阙艳华，董新颖，王娜，等. 中英文对照京剧服饰术语［M］. 覃爱东，陶西雷，校译. 北京：学苑出版社，2017：112，略有改动.

❷ 同❶112.

❸《古代汉语词典》编写组. 古代汉语词典［M］. 北京：商务印书馆，2007：1424.

❹ 霍恩比. 牛津高阶英汉双解词典［M］. 王玉章，赵翠莲，邹晓玲，等译. 7版. 北京：商务印书馆，2010：1781.

❺ 同❶117.

commonly seen.❶

译文剖析 老人指的是上了年纪的或较老的人。京剧中的老人巾主要用于彬彬有礼的长者或慈祥的神仙。查阅《牛津高阶英汉双解词典》，"elderly"的意思是"used as a polite word for 'old'，年级较大的，上了年纪的（婉辞，与old同义）"❷。其内涵远少于"彬彬有礼的长者或慈祥的神仙"，且原文中有关京剧中具体人物使用老人巾的案例也忽略不译，包括样式与学士巾相同的内容也被删节，可见译者的翻译策略是关照要旨、舍弃部分细节。

8. 八卦巾

原文 "八卦巾"样式同高方巾。由缎料制成，前脸正中镶玉帽正，帽身彩绣八卦图和太极图，巾背悬垂两条平金绣飘带，颜色有黑、紫两种。主要为道家、军师所戴。❸

译文 Baguajin (Eight-Diagram kerchief) is one kind of cap with a soft inner lining, embroidered with eight diagrams and Taiji symbols. The crown is flat and the lower part is square. There is a piece of jade, a "cap adjuster" (Maozheng)，and two embroidered ribbons attached to the back. The two colors used are black or purple. It is mostly worn by characters well-versed in Taoism.❸

9. 道巾

原文 "道巾"又称"道士巾"。样式同"八卦巾"，由缎料制成，帽身彩绣阴阳鱼和竖条图案，颜色有黑、紫两种。主要为道士、神仙及足智多谋的军师所戴。❹

译文 Daojin (Toist Kerchief) is one kind of cap with a soft inner lining, similar in design to the Baguajin (Eight-diagram Kerchief). The cap is black or purple, and embroidered with the pattern of eight diagrams. It is worn by Taoist priests and immortals.❹

译文剖析 由于八卦巾和道士巾样式相同，且都与道教相关，故将两者一并剖析。八卦属于中国道家文化，是一套用四组阴阳组成的形而上的哲学符号，用来表示事物自身变化的阴阳系统。"—"代表阳，"- -"代表阴，按照大自然的阴阳变化平行组合形成八种不同形式，称为八卦。查阅《牛津高阶英汉双解词典》，"diagram"的意思是"a simple drawing using lines to explain where sth. is, how sth. works, etc. 简图；图解；图

❶ 阙艳华，董新颖，王娜，等. 中英文对照京剧服饰术语［M］. 覃爱东，陶西雷，校译. 北京：学苑出版社，2017：117，略有改动.

❷ 霍恩比. 牛津高阶英汉双解词典［M］. 王玉章，赵翠莲，邹晓玲，等译. 7版. 北京：商务印书馆，2010：644.

❸ 同❶118，略有改动.

❹ 同❶119，略有改动.

表；示意图"❶。八卦属于中国特有的文化现象，英文中并没有完全对等的词汇，译文 "eight-diagram" 虽然表达出八卦是一种图形，但是并没有完全阐释其特有的文化内涵。道巾中的道源自于中国道教，该头饰在京剧中用来指代道士、神仙及足智多谋的军师。在《牛津高阶英汉双解词典》中，"taoism" 的含义是："a Chinese philosophy based on the writings of Lao-tzu，道家"❷，可见 "taoist" 的译文与道的中文含义基本对等，方便西方读者理解该头饰为中国道教人士所用。但是，该术语的译文忽略了此头饰还适用于神仙或军师，建议在其后的具体描述中，给予进一步的协调，方便服饰文化的交流传播。

10. 方巾

原文 "方巾" 由缎料制成，素面无绣，形似斜坡形屋顶。颜色有黑、蓝、古铜色等，以此来区分人物的年龄和身份。主要为穷生、有功名的书生、品德高尚的贤士所戴。❸

译文 Fangjin (Square Kerchief) is a square cap with a soft inner lining that looks like a sloping roof. Without any embroidery, it has piece of white jade in the middle of the front. Different colors such as black, blue or bronze, are used to distinguish the character's age and status. There are two types: the higher status cap (called Aifangjin) and Gaofangjin. Since actors of Chou (Comic Role) representing scholars always wear the Aifangjin. This type of satin is thus named Fangjin, which is for Qiongsheng (Poor Young Male) role, scholars who have passed county examinations or men of intelligence and integrity. Officials also wear Gaofangjin once they have left office.❹

译文剖析 方巾在京剧中因其形状为正方形而命名。《牛津高阶英汉双解词典》中，"square" 的解释为："having four straight equal sides and four angles of 90"❺。因此，将其直译为 "Square Kerchief" 与原文意思对等。此外，作者还对方巾佩戴者的人物身份地位，以及方巾的分类，包括颜色等都进行了具体说明，方便西方读者深层次了解方巾所蕴含的文化内涵。

11. 鸭尾巾

原文 "鸭尾巾" 由绸缎制成，有软、硬之分。软鸭尾巾，定套，死口，下圆上扇形。

---

❶ 霍恩比. 牛津高阶英汉双解词典［M］. 赵翠莲，译. 8版. 北京：商务印书馆，2014：549.
❷ 同❶2065.
❸ 阚艳华，董新颖，王娜，等. 中英文对照京剧服饰术语［M］. 覃爱东，陶西雷，校译. 北京：学苑出版社，2017：113.
❹ 同❸113，略有改动.
❺ 同❶1951.

由毡帽演化而来，改毡制为绸制。顶部排须放大，如鸭尾，故名。前口水纹，正面镶白玉一块。颜色有古铜、宝蓝、黑色等。古铜色为剧中商贾、店主之冠戴，如《四进士》中的宋世杰、《秦香莲》中的张三阳等；宝蓝色或黑色的如《连升店》中的店家、《断桥》中的许仙等角色佩戴。硬鸭尾巾，两侧缀有绒球、珠子，为老年的江湖人士所戴，如《嘉兴府》中的鲍赐安、《八蜡庙》中的褚彪等。❶

译文 Yaweijin (Ducktail Cap) is one kind of cap that is fixed in size and can not be adjusted. It is now made of satin, having evolved from a felt hat. It is tall narrow in shape, and tilts slightly rippling water, and has white jade to the front. It appears in two main colors: bronze is worn by merchants and shopkeepers, such as Song Shijie in *Four Successful Candidates of the Imperial Examination (Si Jin Shi)* and Zhang Sanyang in the play *Qin Xianglian*; whilst royal blue or black is worn by shopkeepers in *Liansheng Inn (Lian Sheng dian)* and Xu Xian in *Broken Bridge (Duan Qiao)*. Furthermore, there is a hard version of this hat with pompons and pearls on both sides, which is worn by elderly characters. These include Bao Ci'an in *Prefecture of Jiaxing (Jiaxing Fu)* and Chu Biao in the play *Bala Temple (Ba La Miao)* .❶

12. 棒槌巾

原文 "棒槌巾"样式同皇巾，上小下大，前低后高。帽身彩绣花卉、蝴蝶等图案，帽背插一对桃叶形小翅。颜色有绿、白、黄等，使用时需与人物服色相配合。主要为衙内、恶少以及纨绔子弟所戴。❷

译文 Bangchuijin (Wooden Club Kerchief) is one kind of cap with a soft inner lining, made of satin and embroidered with flowers. It is worn by "young masters" or the pampered, carefree offspring of high officials. With the upper section large than the lower section, the cap looks like a wooden club, hence the name. At the back of the cap, a pair of small wings attached to either side of the cap flaps up and down when walking, matching the flippant behavior of the character.❸

13. 荷叶巾

原文 "荷叶巾"又称"四棱巾"，由缎料制成，形似荷叶形，帽顶有檐，帽身彩绣蝙

---

❶ 阙艳华，董新颖，王娜，等. 中英文对照京剧服饰术语［M］. 覃爱东，陶西雷，校译. 北京：学苑出版社，2017：114，略有改动.

❷ 同❶116.

❸ 同❶116，略有改动.

蝠或八宝等图案。主要为丑行（小花脸）扮演的角色所戴。❶

译文 Heyejin (Lotus Leaf Kerchief) is made of green satin, with a soft inner lining, and embroidered with flower designs. It has a square crown with a wide brim in the shape of a lotus leaf on the top. There are also four raised angles on the crown. It is commonly worn by scholars played by the Chou (clown).❶

译文剖析 原文中的蝙蝠或八宝等图案在译文中协调为"flower designs"，忽略了其图纹的文化内涵，可以补充协调为"embroidered with patterns of bats and eight treasures to indicate good luck."

14. 桥梁巾

原文 "桥梁巾"又称"张生巾"，因样式呈桥梁状而得名。由软缎制成，帽身彩绣如意头、花卉等图案，巾背悬垂两条飘带，使用时与人物服色相配合。主要用于较为文雅的年轻书生。❷

译文 Qiaoliangjin (Bridge Kerchief) is so named because its crown is in the shape of a bridge. It is made of satin with a soft inner lining, and embroidered with flower patterns. Two ribbons are attached to the back. Wearing the kerchief suggests that the character is handsome and scholarly. Generally, its color and pattern are determined by the costume.❷

15. 一字巾

原文 "一字巾"样式为一寸宽的长条带，使用时束于角色头部一侧，余带下垂。主要为不束发的童仆、村童使用，如《钓金龟》中的张义。❸

译文 Yizijin (Character "One" Kerchief) is an accessory worn by child servants or village boys, who do not bind their hair up like adults. Made of black satin to width of one cun (Chinese inch, equals y6 decimetre), it has a knot at one side from which the rest of the ribbons dangle. Zhang Yi in *Catching the Golden Tortoise (Diao Jingui)* wears this kerchief.❹

译文剖析 鸭尾巾、棒槌巾、荷叶巾、桥梁巾和一字巾这五种头饰，包括前面的方巾，实际都是以形状命名的，故在此一并剖析其各自译文。"ducktail"属于复合词，由"duck"（鸭子）和"tail"（尾巴）组合而成，作为定语修饰限制其主干词"cap"，而其余的几类巾都直译为"kerchief"。

❶ 阙艳华，董新颖，王娜，等. 中英文对照京剧服饰术语［M］. 覃爱东，陶西雷，校译. 北京：学苑出版社，2017：115，略有改动.

❷ 同❶118，略有改动.

❸ 同❶119.

❹ 同❶119，略有改动.

棒槌巾也是据其形状而命名的。棒槌的意思为木制的工具，圆形，长约67厘米，一端稍粗，便于槌衣；一端较细，便于手握。在《牛津高阶英汉双解词典》中，"wooden" 的释义为："made of wood，木制的，木头的" [1]；"club" 的意思为："a heavy stick with one end thicker than the other, that is used as a weapon，击棍（一头粗，一头细）" [2]。所以棒槌和 "Wooden Club" 在意思上较为对等，且符合这一头饰多为衙内、恶少以及纨绔子弟佩戴的特点。虽然随着时代的发展，棒槌含义也有所不同，除其本义之外，还有傻子或笨蛋之意，但是在京剧服饰的语境中，其名称依然源于形制。

对于荷叶巾（Lotus Leaf Kerchief）、桥梁巾（Bridge Kerchief）以及一字巾（Character "One" Kerchief），其基于原文的直译简洁明了，且在其各自随后的每种头饰的解释中，都分别给予了具体介绍，反映出佩戴者的身份，传达出一定的文化内涵。

综上所述，如果想要凸显不同头饰的形制，不妨补加 "-shaped"，分别翻译为 "Ducktail-shaped Kerchief, Wooden Club-shaped Kerchief, Lotus Leaf-shaped Kerchief, Bridge-shaped Kerchief, Character "One" -shaped Kerchief"，可以在一定程度上更易于西方读者的理解与接受。但是从术语翻译的角度而言，为了后续的解释和说明，且因术语翻译的特殊语境，应该更加简洁，所以音译同样也不失为不同阶段文化协调的策略之一。

16. 大板巾

原文　"大板巾" 亦称 "大页巾" "大披巾"。四顶为一堂，巾口绣 "寿" 字，巾背加平金绣后兜和飘带，使用时与人物服色相配合。主要为校尉、中军和刀斧手、棋牌所戴。若正、反两方龙套同时出场，则正方的龙套戴大板巾，此外其也在巡抚升堂时用。[3]

译文　Dabanjin (Large Board Kerchief), also known as Dayejin (Large Sheet Cap), is a cap with a soft inner lining. The Chinoses seal character "Shou" (longevity) is embroidered on its opening and a hood with gold thread embroidery is attached to its back. It is worn by non-speaking parts such as soldiers, members of the entourage for a marshal or high-ranking official. Walk-on roles on "the right side of history" have the privilege to wear the large board cap. The color should be the same as that of the rest of the costume. Moreover, the cap must be worn by four performers at the same time; thus, four pieces make one set (Tang) . [3]

---

❶ 霍恩比. 牛津高阶英汉双解词典 [M]. 王玉章，赵翠莲，邹晓玲，等译. 7版. 北京：商务印书馆，2010：2316.

❷ 同❶366.

❸ 阙艳华，董新颖，王娜，等. 中英文对照京剧服饰术语 [M]. 覃爱东，陶西雷，校译. 北京：学苑出版社，2017：120.

17. 小板巾

原文 "小板巾"亦称"小页巾""龙套巾"。四顶为一堂，样式同大板巾，但尺寸较小且不带后兜。使用时与人物服色相配合，主要为交战双方的龙套所戴。●

译文 Xiaobanjin (Small Board Kerchief), also called Xiaoyejin (Small Sheet Cap), is one kind of cap with a soft inner lining. It is similar to the Dabanjin (Large Board Cap), but is smaller in size and without a hoop attached to the back. It is also worn by walk-on actors, such as the entourage of a General or Perfect, or by soldiers who wear Shangshouyi (Yellow Soldier's Coat with A Crossover Collar) or Zukan (Soldier's Waistcoat).●

译文剖析 大板巾与小板巾样式相同，适合于不同佩戴者。大板巾为身份地位稍高的人物佩戴，主要是校尉、中军和刀斧手；而小板巾则由交战双方跑龙套的角色使用，没有细化到具体人物。西方读者只能从其英译中注意到两者尺寸大小之差异，而不能由此推断佩戴此头饰人物身份的高低差别。"board"在《牛津高阶英汉双解词典》中的解释为："a long thin piece of strong hard material, especially wood, used, for example, for making floors, building walls and roofs and making boats，板；（尤指）木板"●。原文的大板巾和小板巾中的板直译均为"board"，但是板的本意为质地较硬的木质材料，但是京剧头饰中的巾为软胎，在一定程度上容易误导西方读者认为该巾与木质材料具有关联。因此，不妨依旧在"board"后面补加"-shaped"以凸显其形制，借助协调，方便读者理解接受。

（四）帽

1. 软罗帽

原文 "软罗帽"上半部分是软胎，可以左右、前后、上下晃动或拉歪，分花、素两种。软花罗帽，由缎料制成，帽身彩绣四季花草图案，使用时色彩、花型与人物衣服相配合（戴此帽的人物多穿绣花抱衣抱裤）。主要为富家豢养的武士或江湖豪客所戴，在剧中一般都有激烈的开打场面。软素罗帽，纯黑色，以黑色光缎或黑平绒做面。主要为江湖侠客义士所戴，如《武松打虎》中的武松、《翠屏山》中的石秀等。若向右拉歪、则表示家丁、庄丁一类的人物；若向前拉歪，则表示丑行角色。❸

译文 Ruanluomao (The Soft Silk Hat). The black version is covered with plain satin or velveteen, and is mostly worn by commanders of constables. With soft inner lining in

---

❶ 阙艳华，董新颖，王娜，等. 中英文对照京剧服饰术语［M］. 覃爱东，陶西雷，校译. 北京：学苑出版社，2017：120.

❷ 霍恩比. 牛津高阶英汉双解词典［M］. 王玉章，赵翠莲，邹晓玲，等译. 7版. 北京：商务印书馆，2010：205.

❸ 同❶122，略有改动.

175

upper half, it can be positioned on the head in different ways (askew upward or downward, front or back. to the left or the right). It is also worn by righteous gatekeepers or servants. Soft Luomao can also be classified into two types: embroidered and plain black. The soft embroidered version is made of satin. The colour and pattern need to match the costumes (especially with the embroidered warriors' tight-fitting outfits). It is similar to the hard embroidered silk hat in use. Being mostly worn by warriors from rich families or heroes, who appear in intense fighting scenes. The plain black version is covered with plain-coloured satin or velveteen. It is mostly worn by chivalrous roles. Wu Song in *Wu Song Fights the Tiger* (*Wu Song Da Hu*) and Shi Xiu in *Cuiping Mountain* (*Cui Ping Shan*) are two examples. If the hat is positioned to the right of the head, it means the character is a servant: if it is positioned to the left, it means the character is a comic chou role.[1]

2. 硬罗帽

原文 "硬罗帽"又称"大罗帽",有花、素两种。硬胎花罗帽,帽身硬挺,呈六瓣六角形,每瓣上镶光珠、绒球。主要为江湖侠客或盗贼所戴,如《大破铜网阵》中的白玉堂、《连环套》《骆马湖》中的黄天霸、《除三害》中的周处、《黄隆基》中的黄隆基、《花田错》中的周通等。[2]

译文 Yingluomao (Silk Hat), one kind of hat, is fixed in size and can not be adjusted. It can be classified into two types. The hard Luomao is also known as a big silk hat, which can be further classified into embroidered and plain black versos. Hexagonal in shape, the embroidered version is decorated with pearls and pompons on every petal, and at be peak of the hat there is a round knot. There are two types of embroidered hat: one covered with coloured embroidered satin and the other covered with gold adornments over wire guaze and with silver decorations inlaid in satin. This type of hat is most used by knight-errants or burglars. Examples include Bai Yutang in *The Brass Net Plan* (*Tong Wang Zhen*), Huang Tianba in *Strategy of lmterlocking Rings* (*Lian Huan Tao*) and *Luoma Lake* (*Luo Ma Hu*), Zhou Chu in *Exorcising Evil* (*Chu San Hai*), Huang Longji in *Huang Longi* and Zhou Tong in *The Mistakes at the Flower Festival* (*Hua Tian Cuo*), etc.[3]

原文 The Luomao is a six-sided fabric hat mounted on a tall circular band that fits around

[1] 阙艳华,董新颖,王娜,等. 中英文对照京剧服饰术语 [M]. 覃爱东,陶西雷,校译. 北京:学苑出版社,2017:122,略有改动.
[2] 同[1]169.
[3] 同[1]169,略有改动.

the crown of the head. This hexagonal beret-like hat is either stiff and worn upright, Yingluomao, or soft and worn collapsed to one side, Ruanluomao. It has a padded ball at the apex of the six pieces. This hat can be either plain or embroidered, and the use changes depending on the version. A solidcolor hat is called Suluomao (plain Luomao). Head servants or attendants wear a black Suluomao with the top standing up, and a black Suluomao is worn tilted to the right by unattached soldiers wearing the black Kuaiyi (Martial Clothing). When worn with the embroidered Baoyi (Hero Jacket and Trousers), the Luomao is made with matching fabric and embroidery patterns.❶

拙译 罗帽的材质为纺织品，有六个面，底部圆形可以适合头部的大小尺寸。硬罗帽为六角，类似贝雷帽，直挺；软罗帽柔软，可对叠。帽子六瓣的顶端最终汇聚到头顶上带衬垫的小球。该帽有花、素之分，依款式变化。纯色罗帽是素罗帽。总管或侍者头戴黑色素罗帽，身穿黑色快衣；当身着刺绣的抱衣（英雄衣裤）时，罗帽的材质和刺绣图案需要与之匹配。

译文剖析 罗帽分为软胎和硬胎两种。硬胎为硬罗帽，软胎则为软罗帽。查阅《牛津高阶英汉双解词典》，"silk"被解释为"a type of fine smooth cloth made from silk thread，丝织物；丝绸"❷。根据罗帽的质地不同，硬罗帽的协调翻译为"Silk Hat"，没有凸显汉字的硬；软罗帽的协调翻译为"The Soft Silk Hat"。在《牛津高阶英汉双解词典》中"soft"的解释为"changing shape easily when pressed; not stiff or firm，软的；柔软的"❸，"soft"强调了罗帽的质地为软胎，并与硬罗帽的翻译"Silk Hat"形成对照，方便西方读者理解两者的差异与共性。

笔者的拙译也同样在句式结构、词汇选用、语序调整等方面充分考虑到中西语言的文化差异，协调源语与译语之间的差异，充分体现了汉语凝练简洁的特点，方便读者理解与接受。

3. 梢子帽

原文 "梢子帽"又称"骚子帽""烟毡帽"。样式呈圆形，檐向外翻。由白色毛毡制成，周围镶黑边或蓝边。戴时，捏成扁形，斜戴于头上。主要为剧中解差、老军、士兵等人物所戴。❹

译文 Shaozimao (Sentry Post Cap) is a decorative felt cap in the shape of an inverted boat, edged with a black brim, and worn by soliders or those who escort criminals.❺

❶ BONDS ALEXANDRA B. Beijing Opera Costumes: the Visual Communication of Character and Culture [M]. Honolulu: University of Hawaii Press，2008：257.

❷ 霍恩比. 牛津高阶英汉双解词典 [M]. 7版. 王玉章，赵翠莲，邹晓玲，等译. 北京：商务印书馆，2010：1867.

❸ 同❷1911.

❹ 阙艳华，董新颖，王娜，等. 中英文对照京剧服饰术语 [M]. 覃爱东，陶西雷，校译. 北京：学苑出版社，2017：123.

❺ 同❹123，略有改动.

4. 毡帽

原文 "毡帽"由毛毡制成。样式呈圆锥形，顶端有一片排须，帽口向外翻。颜色有红、蓝、白等。红色毡帽又称"素椒帽"，四顶为一堂，主要为三班衙役或更夫所戴，如《四进士》中的衙役。蓝色毡帽主要为年轻庶民、渔夫、店伙、车夫、堂倌所戴。白色毡帽，主要为老翁、苍头、乡邻所戴。❶

译文 Zhanmao (Felt Cap) is a decorative round felt cap in red, blue or white. The blue felt cap is used for younger male characters, fishermen, shopkeepers, drivers and by grooms. White felt caps are used to represent old men, soldiers and neighbours.❶

原文 The felt hat has a round crown and upturned brim. It comes in white colors and is used for servants. The boatman's version is called Shaozimao.❷

拙译 毡帽样式呈圆形、檐向外翻，颜色为白色，适用于侍者。船工用的款式称作梢子帽。

5. 鬃帽

原文 "鬃帽"又称"蟋蟀罩""蛐蛐罩"。由黑色马鬃编织而成，为净、丑两行所扮演的绿林豪杰所戴，如《水浒传》中的李逵、刘唐等。又有硬花鬃帽，是在鬃帽上加一鬃帽套。鬃帽套用铁丝编成，呈圆柱形，四周镶饰贴绢蝴蝶，通体缀白色光珠与蓝白色绒球，如秦仁、朱光祖、杨香武等人所戴。❸

译文 Zongmao (Palm Hat) is one kind of black netted hat, made of palm bark, and is round at the bottom and pointed at the top. Resembling the shape of a hood. It also looks like a cricket's nest and is used by painted face Jing and comic Chou characters who are like Robin Hood in disposition, such as Li Kui or Liu Tang in *The Water Margin*. There is also another kind of hard helmet with an additional cylindrical wirecover patched with silk butterflies and decorated with white pearls and blue-white pompons. This is worn by Wuchou martial clown roles, such as Qin Ren, Zhu Guangzu and Yang Xiangwu etc.❹

原文 Headdress and Accessories for Hualian Huaxuezi. A helmet style of headdress can be

---

❶ 阙艳华，董新颖，王娜，等. 中英文对照京剧服饰术语 [M]. 覃爱东，陶西雷，校译. 北京：学苑出版社，2017：124，略有改动.

❷ BONDS ALEXANDRA B. Beijing Opera costumes: the visual communication of character and culture [M]. Honolulu：University of Hawaii Press，2008：257.

❸ 同❶173，略有改动.

❹ 同❶173.

worn with these Xuezi, or a Zongmao (lit."mane hat"), a tall truncated cone of sheer black mesh. With the Hualian Huaxuezi worn open, the characters wear a robe such as the Jianyi, a padded vest, and high-soled, high-topped boots.❶

拙译 花脸花褶子的头饰及配饰。头盔式的头衣，或者鬃帽可以与这些褶子一起穿戴，截面是锥形，呈透明黑色网眼。扮花脸穿花褶子时，京剧人物通常都会穿诸如箭衣的袍服、衬垫背心和高帮靴。

译文剖析 鬃帽和毡帽均是据其制作材质而命名，鬃帽是由黑色马鬃编织而成，故命名为鬃帽，译为"Palm Hat"。查阅《牛津高阶英汉双解词典》，"palm"的意思是"a straight tree with a mass of long leaves at the top, growing in tropical countries，棕榈树"❷，因此"palm"并不等同于马鬃，此鬃非彼鬃。毡帽的译文为"Felt Hat"，《牛津高阶英汉双解词典》中"felt"的意思是"a type of soft thick cloth made from wool or hair that has been pressed tightly together，毛毡"❸，该译文方便西方读者通过辨认帽子的材质理解其具体所指，如此协调的译文较好地传递了原文的本意。

6. 皂隶帽

原文 "皂隶帽"样式呈长方形，由黑缎蒙面，正面彩线绣长形"寿"字，右侧绣云福图案，一侧插孔雀翎。主要为皂隶专用。❹

译文 Zaolimao (Cap of a Yamen Runner) is a decorative rectangular black satin cap with the Chinese character "Shou" (longevity) on the front and peacock feathers to one side. It is worn by runners who work for the government office (Yamen) in feudal China.❹

译文剖析 皂隶是中国古代衙门里社会地位低下的差役，皂隶帽是该类衙役的头衣，皂隶帽英译为"Cap of a Yamen Runner"。但是衙门一词在英文中并没有对应表达，属于汉语特色词汇。拼音"Yamen"在忠实传递原文信息的同时，并未充分考虑到西方读者的接受和理解程度。但是在其后补加的"runner"一词可以在一定程度上帮助西方读者理解其角色身份。正如《牛津高阶英汉双解词典》中"runner"的意思为"a person in a company or an organization whose job is to take messages, documents, etc. from one place to another，信人；信差"❺。虽如此，但是实际京剧中的皂隶属于勤杂或苦役等卑贱工作，与"runner"的信差相比，在字面意思上并不对等，但有"异曲同工"之

❶ BONDS ALEXANDRA B. Beijing Opera Costumes: the Visual Communication of Character and Culture［M］. Honolulu：University of Hawaii Press，2008：161.

❷ 霍恩比. 牛津高阶英汉双解词典［M］. 王玉章，赵翠莲，邹晓玲，等译. 7版. 北京：商务印书馆，2010：1438.

❸ 同❷741.

❹ 阙艳华，董新颖，王娜，等. 中英文对照京剧服饰术语［M］. 覃爱东，陶西雷，校译. 北京：学苑出版社，2017：125，略有改动.

❺ 同❷1752.

京剧服饰三衣类文化内涵及翻译协调美

第五章

妙，在一定程度上促进了京剧服饰文化交流。

7. 风帽

原文 由绸缎制面，软胎，向上收成尖顶状，下端披在肩上。前口上翻，镶饰二寸异色宽边，左右有飘带。人物罩于所戴的冠帽外，表示防风御寒。若用飘带将后兜齐根束住，前额加饰面牌，则表示行路。如《十三妹》中的何玉凤、《捉放曹》中的曹操。❶

译文 Fengmao (literally "Wind Hat") is a decorative satin hat with a long back edge that extends to the shoulders. It is worn with other hats to signify that the character is sheltering themselves from the cold and the driving wind. The Emperor wears a special coloring of this hat in yellow. The female version is richly embroidered. Examples can be found in He Yufeng in the play *Thirteenth Sister* (*Shi San Mei*) and Cao Cao in *The Capture and Release of Cao Cao* (*Zhuo Fang Cao*).❷

译文剖析 风帽属于京剧头饰中比较特殊的一种，佩戴时罩在所戴冠帽外，用来防风御寒。风帽协调翻译为 "Wind Hat"，用词简洁且能体现原文本意，便于西方读者理解接受。

8. 毗卢帽

原文 "毗卢帽" 又称 "僧帽"。样式呈椭圆形，上宽下窄，红缎蒙面加黄边，翻檐，前面正中绣 "佛" 字，四周镶莲花瓣。如《三打白骨精》中的唐僧、《白蛇传》中的法海、《游六殿》中的目连僧等。若帽前附加五佛冠，左、右饰飘带，则为禅师施行法术的特定标识。❸

译文 Pilumao (Buddhist Head Monks Crown) is also known as "Sengmao" ("seng" means "monk") and is one kind of crown exclusively worn by monks and priests. Oval in shape, it is covered with red satin and has a yellow brim with upturned eaves. With the Chinese character "Fo" (means "Buddha") embroidered on the front, it is decorated all over with lotus petals. Characters that use this crown include the Tang Monk in *Monkey King Subdues White Bone Demon* (*San Da Baigujing*), Fa Hai in *The White Snake* (*Bai She Zhuan*) and the Mulian Monk in the play *Suffering in Hell* (*You Liu Dian*). If the hat has a Wufoguan (Five Buddha Crown), attached to the front, with ribbons hanging down from the two sides, it means the wearer is in the process of casting a spell.❹

---

❶ 阚艳华，董新颖，王娜，等. 中英文对照京剧服饰术语［M］. 覃爱东，陶西雷，校译. 北京：学苑出版社，2017：125.
❷ 同❶125，略有改动.
❸ 同❶126.
❹ 同❶126，略有改动.

9. 男僧帽

原文 "男僧帽"又称"和尚帽"。帽前身高四寸半、帽后身较低，前面正中绣"佛"字。分软、硬两种：软和尚帽为小和尚佩戴，如《蜈蚣岭》中的小和尚；硬和尚帽为大和尚或者主要角色佩戴。❶

译文 Nansengmao (Cap for a Male Buddhist Monk)is a decorative cap in black or bromine color, with the Chinese character "Fo" (means "Buddha") on the front. It is worn by both young and old monks.❷

原文 The Sengmao (Monk Hat) is peaked in front and has upturned flaps above the ear area. The character for Buddha is embroidered above the forehead.❸

拙译 僧帽前部高耸，耳部上方帽檐翻转。帽额有"如来佛"字样的刺绣。

10. 女僧帽

原文 "女僧帽"又名"尼姑帽"。样式呈圆形。平顶，顶上抽褶，帽前镶白玉帽正，颜色有黑、灰两种。主要为女尼专用。❹

译文 NüSengmao (Cap of a Buddhist Nun) is a decorative round cap with a flat tog and jade inlay. It is either black or grey in color to distinguish the age of the character. It is specially worn by Buddhist nuns.❷

译文剖析 从8到10这三顶头衣都是针对僧帽的不同描述，译文8基于原文内容，包括结合戏曲《三打白骨精》《白蛇传》以及《游六殿》中的人物等全部给予翻译。而译文9则将类似的"《白蛇传》中的法海"省译，同时基于原文本意进行了拙译，以协调句式结构，将原文的两句协调翻译为有停顿的切分句式，通过逗号拆开原句，方便读者阅读理解；其中文化特色词汇"Buddha"被协调翻译为如来佛，符合了汉语阅读的需求和习惯。译文10则忠实于原文句意，整合原文句式结构，关照英汉语言句式差异，包括特色文化词汇"Buddhist nuns"与女尼之间的转换，较为忠实地传递了原文的文化内涵。

---

❶ 阙艳华，董新颖，王娜，等. 中英文对照京剧服饰术语 [M]. 覃爱东，陶西雷，校译. 北京：学苑出版社，2017：127.

❷ 同❶127，略有改动.

❸ BONDS ALEXANDRA B. Beijing Opera Costumes: the Visual Communication of Character and Culture [M]. Honolulu: University of Hawaii Press，2008：198.

❹ 同❶126.

京剧服饰三衣类文化内涵及翻译协调美

第五章

11. 红缨帽

原文 "红缨帽"又称"小钿帽"。样式呈圆形，上宽下窄，四周翻檐。由黑缎或黑皮绒制成，顶部镶有红、蓝色宝珠，宝珠下端饰红色丝穗，主要为番邦人物或清朝官员专用。❶

译文 Hongyingmao (Tartar Cap) is a decorative black satin hat, edged with feather or wool, worn by all the nomadic peoples in the north of ancient China.❷

译文剖析 红缨为红色线或绳制成的饰品。红缨帽的帽顶披红缨，作为清代礼帽，为番邦人物或清朝官员专用。查阅《牛津高阶英汉双解词典》"tartar"解释为"a person in a position of authority who is very bad-tempered，暴君；脾气暴躁的掌权者"❸。"Tartar Cap"在指代当权者的同时，也暗示其穿用者是"脾气暴躁的掌权者"。如此译文在一定程度上容易误导读者，"tartar"与红缨帽实际佩戴人物的性格之间存在文化差异，追根溯源，其本质差异在于东西方对于不同色彩的联想意义存在差异，西方的红色始终与暴力或流血相关。

12. 砂锅浅儿

原文 "砂锅浅儿"属于圆顶帽式，帽浅如砂锅，故得名。由黑色绉绸或平绒制面。主要为高级官府中门吏、仆役所戴，如《甘露寺》中的乔福、《杨家将》中的杨洪等。戴时，加饰一道橙色或蓝色的包头条，一半露于帽檐外，以示年迈。❹

译文 Shaguoqian'er (Casserole Hat) is a domed hat that resembles a casserole, hence the name. It is covered with black crepe or velveteen, and is worn by gatekeepers or servants in senior positions in governments. Qiao Fu in *Ganlu Temple* (*Gan Lu Si*) and Yang Hong in *Generals of the Yang Family* (*Yang Jia Jiang*) are two examples of characters that wear this costume. Attaching a strip of orange or blue head-cloth from the brim of the hat shows that the character is elderly.❷

译文剖析 原文中砂锅浅儿的字面意思为较浅的砂锅，实指京剧中一种形如砂锅的帽子，其帽膛较浅。在《牛津高阶英汉双解词典》中，"casserole"的解释为"a container with a lid used for cooking meat, etc. in liquid in an oven，炖锅；砂锅"❺。译文选用英文中对应的词汇，协调源语与译入语的差异，形象再现了该帽的款式，便于西方读者的理解与接受。

---

❶ 阚艳华，董新颖，王娜，等. 中英文对照京剧服饰术语 [M]. 覃爱东，陶西雷，校译. 北京：学苑出版社，2017：128.

❷ 同❶128，略有改动.

❸ 霍恩比. 牛津高阶英汉双解词典 [M]. 王玉章，赵翠莲，邹晓玲，等译. 7版. 北京：商务印书馆，2010：2068.

❹ 同❶127，略有改动.

❺ 同❸297.

13. 皇帽

原文 "皇帽"又称"王帽""唐帽""堂帽"。前扇微圆，正中有绒球面牌，顶端有黄色大绒球两个，旁缀珠子。左右两侧装龙形耳子，下垂流苏。后扇椭圆形，有朝天翅一对。全帽前低后高，黑底贴金点绢，主色为金，而附饰的绒球、流苏颜色不同，黄色多于红色。如《斩黄袍》中的赵匡胤❶和《甘露寺》中的刘备。❷

译文 HANG MAO❸ (Emperors' Crown) is a crown exclusively worn by emperors. The crown is fixed and can not be altered. The front of the crown sits lower than the back, and is decorated with the pattern of nine dragons. The front fan-shaped oval is decorated with two large decorative yellow balls on the top and dozens of beads on both sides. Set with the decoration of a dragon's head and body on both sides. It forms a dragon-like tail near the ears, from which tassels dangle. The main colour is yellow, and the decorative balls attached to it come in different colors. The gold crown trimmed with different decorative balls and tassels is worn by such roles as Zhao kuangyin in *Slash the Impenal Robe* (*Zhan Huang Pao*), who are representative of "orthodox" emperors. The crown trimmed with decorative balls and red tassels is worn by such roles as Liu Bei in *Ganlu Temple* (*Gan Lu Si*), who are representative of unorthodox emperors.❹

译文剖析 从译文与原文字数差异的角度，可见译文基于原文几乎没有任何删节，在句式安排方面也几乎遵照原文顺序。但是较为凸显的是原文中有关"皇帽"的两个案例，即《斩黄袍》中的赵匡胤和《甘露寺》中的刘备，对于西方读者相对并不熟悉，所以译文适度补充了相关信息，以方便读者理解。

14. 鞑帽

原文 "鞑帽"又称"双龙鞑帽"，为黄缎所制的金底翻边圆帽，上铸双龙，故名。帽顶镶玉饰，旁缀珠子、小绒球，正中面牌缀橙色大线球。帽后垂挂两条橙色绣龙飘带和橙色丝穗。分橘黄色和明黄色两种，明黄色为皇帝微服出行时所戴，橘黄色为王侯将帅所戴。❺

译文 Damao (Cuffed Round Hat), also known as a flying-dragon hat, is a cuffed round hat with a gold bottom, and is worn by emperors when they are travelling incognito,

---

❶ 原文为"《贺后骂殿》中的赵光义"，为避免歧义，改为"《斩黄袍》中的赵匡胤"。
❷ 阙艳华，董新颖，王娜，等. 中英文对照京剧服饰术语 [M]. 覃爱东，陶西雷，校译. 北京：学苑出版社，2017：129.
❸ 经查证，此处为拼写错误，应改为"Huangmao"。
❹ 同❷129，略有改动.
❺ 同❷132.

or generals from foreign lands. It is made of yellow satin and cast with two orange dragons, inlaid with jade ornaments on the top, and is set with dozens of beads and small decorative balls on both sides, with a big decorative orange ball in the middle of the Mianpai (a circular ornament used in costumes for acrobatic fighting roles). Embroidered orange ribbons and orange tassels dangle from the back section of the hat.❶

译文剖析 鞑是对我国古代北方各游牧民族的统称，鞑帽专用于特定王侯将相的行路场合。查阅《牛津高阶英汉双解词典》，"cuff" 的解释为 "the end of a coat or shirt sleeve at the wrist，袖口"❷；"round" 的解释为："shaped like a circle or a ball，圆形的；环形的；球形的"❸；"hat" 的意思是 "a covering made to fit the head, often with a BRIM, and worn out of doors,（常指带檐的）帽子"❹。鞑帽协调翻译为 "cuffed round hat"，充分考虑到帽子的形状，但是无法顾及鞑的具体所指，在一定程度上造成了文化的损耗，无法再现原文的"原汁原味"。

15. 文阳帽

原文 "文阳帽"又称"汾阳帽"。外形基本与相貌相同，但尺寸较大略高。有金胎、银胎两种，上缀绒球、珠子。前扇正中饰面牌，左右对称镶饰双龙，成二龙戏珠图案。后扇竖插双龙绞柱立翅，下横插一对金色平面镂龙如意翅。主要为权势显赫的宰相或有大功勋的勾脸角色专用，如《草桥关》中的姚期、《逍遥津》中的曹操、《打严嵩》中的严嵩。❺

译文 Wenyang (Golden Marten) is one kind of golden wire gauze cap worn on official occasions. It is similar to the "Premier's Marte" in shape, but it is larger in size and is slightly taller. The cap has an inner lining made of wire gauze, which is gold with a dragon design, and further decorated with pompons and pearls. It is embellished with a big pompon and white jade positioned in the centre of the front, inlaid with two symmetrical dragons to form a pattern of two dragons chasing a pearl. At the back, there is a pair of upright wings adored with two dragons twisting a pillar, and a pair of golden "Ruyi" wings decorated with impressed dragons at the bottom. The headdress is for powerful auxiliary officials or meritorious painted-face roles, such as Yao Qi in the play *Caoqiao Pass* (*Cao Qiao Guan*), Cao Cao in *Xiaoyao Ferry* (*Xiao Yao Jin*) and Yan Song in *Fighting Yan Song* (*Da Yan Song*). Guo Ziyi who is the Prince of Fenyang in *Taming of the Prince* (*Da Jin Zhi*), wears it .

❶ 阙艳华，董新颖，王娜，等. 中英文对照京剧服饰术语［M］. 覃爱东，陶西雷，校译. 北京：学苑出版社，2017：132.
❷ 霍恩比. 牛津高阶英汉双解词典［M］. 王玉章，赵翠莲，邹晓玲，等译. 7版. 北京：商务印书馆，2010：486.
❸ 同❷1740.
❹ 同❷935.
❺ 同❶134.

Hence，it is also known as the "Fenyang Cap".**❶**

译文剖析 文阳帽又称汾阳帽，因角色 "汾阳王" 爵号而得名。在中国古代唐朝时期，郭子仪因平息 "安史之乱"，收复洛阳、长安两地有功，被奉为 "汾阳王"。汾阳帽为金色，与黑相貂相对，故又称金貂。查阅《牛津高阶英汉双解词典》，"golden" 的意思为 "bright yellow in color like gold，金色的；金黄色的"**❷**，表明了 "头衣" 的颜色。此外，"marten" 在《牛津高阶英汉双解词典》中的解释为 "a small wild animal with a long body, short legs and sharp teeth，貂"**❸**。貂属于兽类，其皮毛也常作为头衣的材质。因文阳一词在英文中并没有对应的词汇，译者只好根据其颜色及材质将其协调翻译为 "Golden Marten"，但是对文阳文化内涵的阐述只能在随后的译文中将其与官位联系。作为术语的文化翻译，往往因为篇幅与语境等因素的限制，文化的侧重与流失在一定程度上确实无法避免。

16. 侯帽

原文 "侯帽" 又称 "耳不闻"。样式呈平顶方形，两侧有形同翼状的宽翅，宽翅四周垂黄色短穗。前额正中饰大线球面牌，左右有龙头、光珠。有金胎、银胎两种。主要为封侯者所戴，如《大保国》中的徐彦昭、《敬德装疯》中的尉迟恭等。另外，侯顶加饰戟头，称为 "台顶"，主要为征战疆场、掌握兵权的人物所戴，如《临潼关》中的徐达、《双阳公主》中的杨宗保等。**❹**

译文 Houmao (Marquis Hat) is also known as "Erbuwen" (literally close one's ears). It is one kind of square hat with a flat top. There are two wide floppy wings dropping down on both sides covering up the ears, hence the name. It is commonly worn by founding fathers of a royal clan who are loyal and demonstrate integrity of spirit turning a deaf ear to slander. It is adorned with a decoration in the shape of a flame and has a large pompon in centre of the front section. On the right and left, dragon head-shaped decorations and pearls form the pattern of dragons in the clouds. Yellow tassels drop down around the wide wings. They are two colors for this hat: gold and silver. The hat is normally worn by Marquis. Examples include Xu Yanzhao in *Defending the Nation* (*Da Bao Guo*), Yuchi Gong and Cheng Yaojin in *Jing De Feigns Madness* (*Jing De Zhuang Feng*). If the cap has an additional trident halberd situated on the top, it is known as Taiding and is worn by characters fighting on the battlefield or those who hold military power. Xu Da in *Lintong Pass* (*Lintong Guan*) and Yang Zongbao

---

❶ 阙艳华，董新颖，王娜，等. 中英文对照京剧服饰术语［M］. 覃爱东，陶西雷，校译. 北京：学苑出版社，2017：134.
❷ 霍恩比. 牛津高阶英汉双解词典［M］. 赵翠莲，译. 8版. 北京：商务印书馆，2014：877.
❸ 同**❷**1237.
❹ 同**❶**139.

in *Princess Shuangyang* are two examples.❶

原文　The Houmao is square, and the two sides of the square slope down into wide ear flaps edged in fringe. A single pompon is perched on top of the crown and a long life symbol is placed over the forehead. This headdress is nicknamed "unable to hear" (Erbuwen) because the broad stiff flaps covering the ears give the impression that the wearer does not want to listen to intrigue. The style derives from a story of a real general, Guo Ziyi, a famous leader who refused to listen to the advice of suspected traitors and covered his ears with his hands. When a horn-shaped decoration is added to the top of the headdress, the combination is called a Taiding. The horn, because it is shaped like a weapon, indicates the wearer's military status and therefore it is usually worn by a noble who has a military position.❷

拙译　侯帽又称耳不闻，两侧有形同翼状的宽翅，侯顶加饰戴头，因为耳朵上方宽大的硬皮瓣给人的印象是"佩戴者不想听阴谋"。该帽源自唐朝将军郭子仪，他拒绝听取叛逆者的建议，用手捂住耳朵。当侯帽顶部加上角形装饰时，称作台顶。其形为号角，酷似武器，表示佩戴者的军事地位，通常由掌管军事的贵族佩戴。

译文剖析　侯帽又称为耳不闻，寓"不听谗言"之意，主要为忠心报国的开国元勋及封侯者所戴，故得名侯帽，译者将其翻译为"Marquis Hat"。"marquis"在《牛津高阶英汉双解词典》中的解释为"(in some European countries but not Britain) a NOBELMAN of high rank between a COUNT or a DUKE，（除英国外一些欧洲国家的）侯爵"❸，协调对应于中国古代中的王侯将相，较好地传递了其文化内涵。

17.　纱帽

原文　"纱帽"又称"乌纱帽"。样式分为上下两层，前低后高，帽身蒙黑纱或平绒，清素无花。帽背横插一对乌纱帽翅，帽翅有方、圆、尖等几种，根据不同人物分别使用。生角演文官，多插方帽翅；净角演武官，多插尖帽翅；丑角扮演职位较低的知县或相府的门客，多插圆帽翅。还有桃叶式帽翅，称"学士帽"，如《醉写》中的李白。纱帽还可以插翎挂尾，称"翎尾纱帽"，表示胡服冠戴，如《汉明妃》中的王龙。❹

译文　Shamao (Gauze Cap) is an adjustable cap for officials, who were formally known as "Futou", and commonly known as "Shamao". The body of the cap, known as the

---

❶ 阙艳华，董新颖，王娜，等. 中英文对照京剧服饰术语［M］. 覃爱东，陶西雷，校译. 北京：学苑出版社，2017：139.

❷ BONDS ALEXANDRA B. Beijing Opera Costumes: the Visual Communication of Character and Culture［M］. Honolulu: University of Hawaii Press，2008：255.

❸ 霍恩比. 牛津高阶英汉双解词典［M］. 王玉章，赵翠莲，邹晓玲，等译. 7版. 北京：商务印书馆，2010：1236.

❹ 同❶160.

"Shamaoke" (literally the "shell of the Shamao"), has two sections: the lower front section is semicircular in shape, whilst the higher back section is more rigid circular in shape. It is plain, covered with black crepe or velveteen, and decorated with a pair of upright wings placed towards the rear bottom of the cap. Because of its colour, it is also known as "Wushamao" (Black Gauze Cap). It can be altered by attaching accessories according to the needs of the role: square wings can be added for male Sheng roles when they portray civil officials, pointed wings for painted face Jing roles when they play civil officials, and round wings can be added for comic Chou roles when they act as civil officials. Peach leaf wings for scholars known as "Xueshimao" (Batchelor's Cap), can be used by characters such as Li Bai in *Writing Poems While Intoxicated* (*Zui Xie*). The body of the cap ("Shamaoke") can be used on its own without the wings. It is also worn by Wang Long in the play *Imperial Concubine Ming of Han Dynasty* (*Han Ming Fei*), where it is also known as the "Gauze Cap with Pheasant Feathers and Foxtails" to represent the fact that the character is from an ethnic minority.❶

原文 The Shamao is a general category of hats that have the same basic double-crowned shape with wings. They vary in the materials used, the amount of decoration, and the shape of the wings. The lower front and the raised back crowns are built as separate pieces, to allow for multiple head sizes. Worn by many officials of the traditional Jingju court, a version of the Shamao is often paired with the Mang, the Guanyi, and the Pi (Formal Robe). While the rank of the wearer could be determined by the wings in the past, onstage the wing shape reflects personality characteristics of the wearer. The Shamao may be worn at home as well as for official events.❷

拙译 纱帽的形状基本呈双冠状，是带有帽翅的帽子的总称。其材料、装饰的数量和帽翅的形状各异。纱帽前部较低，后部凸起，彼此独立，以适应头大头小之需。在京剧传统的朝廷场景中，许多官员都头戴纱帽，并常与蟒、官衣和帔（常服）相互搭配。展翅在过去暗示人物的身份等级；不仅如此，在舞台上，展翅的形状还反映出人物的性格特征。纱帽既可以在京剧舞台的惯常表演中选用，也可以在官方的特定活动中穿戴。

18. 中展纱帽

原文 "中展纱帽"又称"方翅纱帽"。展翅呈长方形，外缘略带椭圆，展面贴纱，镂

❶ 阙艳华，董新颖，王娜，等. 中英文对照京剧服饰术语［M］. 覃爱东，陶西雷，校译. 北京：学苑出版社，2017：160，略有改动.

❷ BONDS ALEXANDRA B. Beijing Opera Costumes: the Visual Communication of Character and Culture［M］. Honolulu: University of Hawaii Press，2008：254.

太阳卧水或金龙立水图案，多用于俊扮的老生、小生角色或正派的官员。旧时曾以此帽表示人物忠义，有"忠纱"之名。戴此帽者，如《四郎探母》中的杨延辉、《清风亭》中的张继保、《四进士》中的田伦等。若在中展纱帽上附饰宫花，即为新科状元、进士的特定冠帽，如《御碑亭》中的王有道。若在中展纱帽上附饰长五尺的白色绸条，则为官员挂孝标志，如《朱痕记》中的朱春登。❶

译文　Zhongzhanshamao (Medium Wing Gauze Cap) is also known as "Fangchishamao" ("Fangchi" means "square wings", so the "Square Winged Gauze Cap"). It is a kind of adjustable gauze cap for Laosheng (Old Male Roles) and Xiaosheng(Young Male Roles) of decent character or upstanding officials. The cap's wings are rectanglular with a slightly oval brim, covered with gauze and adorned with patterns of "the sun reflected in sea water" or "golden dragons rising from the sea water", or the character "Shou" (longevity). In old times, it was known as "Zhongsha" (literally "Loyal Gauze [Hat]") usually worn by the most devoted and faithful roles. Consequently, it is used by a large variety of characters, such as Yang Yanhui in the play *Yang Yanhui visits His Mother* (*Si Lang Tan Mu*), Zhang Jibao in the play *Breeze Pavilion* (*Qing Feng Ting*), and Tian Lun in the play *Four Successful Candidates of the Imperial Examination* (*Si Jin Shi*). Furthermore,various embellishments are used to distinguish between different roles. For instance, it is exclusively worn with flowers by the Zhuangyuan (the top winner in the imperial examinations), such as Wang Youdao in the play *Pavilion of the Imperial Tablet* (*Yu Bei Ting*); or with white silk streamers that are five foot long by officials in mourning, such as Zhu Chundeng in the play *The Story of the Red Scar* (*Zhu Hen Ji*).❷

原文　The Zhongsha (Loyal Hat), a style of Shamao in black velvet with the double crown and wings (Chi), is often worn by the Sheng roles who wear Mang. This combination reflects a Ming precedent, when the Shamao was worn with official garments.❸

拙译　忠纱是黑绒双翅帽，通常是穿蟒服的生角儿选用，这种服饰组合由明朝开创，在当时忠纱帽常为官员所戴。

原文　One version of the Shamao is the Zhongsha (Loyal Hat), which is widely used for Laosheng and Xiaosheng characters. A small square of white jade (Maozheng) on the front of the hat indicates a Xiaosheng. The hats worn by both Laosheng and Xiaosheng are made with black

❶ 阙艳华，董新颖，王娜，等. 中英文对照京剧服饰术语［M］. 覃爱东，陶西雷，校译. 北京：学苑出版社，2017：161.

❷ 同❶161，略有改动.

❸ BONDS ALEXANDRA B. Beijing Opera Costumes: the Visual Communication of Character and Culture［M］. Honolulu: University of Hawaii Press，2008：141.

velvet fabric stretched over a firm foundation. The wings on the Zhongsha are flat, upright, and made from golden filigree with wired edges. Upright, wide-rounded trapezoidal wings are reserved for intelligent and loyal court officials. ❶

拙译 纱帽中的忠纱（忠诚帽），常用于老生和小生的角色。帽子前面镶有白玉（帽正），暗示人物身份是小生。老生和小生所戴的帽子通常采用黑天鹅绒面料，配有硬质的帽胎。忠纱的两个展翅平且直，用金线装饰，边饰金属丝。直立、圆润的梯形展翅常为忠诚智慧的朝廷官员所用。

19. 圆展纱帽

原文 "圆展纱帽"又称"圆纱""丑纱"。展翅呈圆形，展面贴金点绢，镂古钱或圆"寿"字图案。展柄的弹簧圈可加长，随人物动作摇摆颤动。主要为昏庸糊涂的贪官，或是谄上骄下、吹牛拍马的势利小人以及奸诈狡猾、阴谋害人的人物所戴，如《审头刺汤》中的汤勤、《荒山泪》中的胡泰来、《失印救火》中的金祥瑞等。另外，一些品级低下的小吏也戴此帽，如《清官册》中的驿丞、《黄金台》中的门官等。❷

译文 Yuanzhan Shamao (Round-winged Gauze Cap) is also known as the Yuanchi Shamao ("Yuanchi" also means "round wings"). This is a gauze cap used exclusively by Chou or comic roles. The wings feature golden decorations inset with satin, and can be decorated with patterns in the form of ancient copper coins, or with the Chinese character "Shou" (longevity). Two extended and sprung coils are added so that the round wings of the cap shake and wave with the movement of the actor. This kind of gauze cap is mostly worn by fatuous muddle-headed and officials, snobs who are good at boasting and fattering, or by insidious characters. These include Tang Qin in *The Fake Head and the Assassination of Tang Qin* (*Shen Tou Ci Tang*), Hu Tailai in *The Miserable Life* (*Huang Shan Lei*) and Jin Xiangrui in *Loss and Recovery of the Official Seal* (*Shi Yin Jiu Huo*) etc. Furthermore, it is also used by some lower-rank minor, officials, such as the postmaster in *List of Incorrupt officials* (*Qing Guan Ce*) and the Doorman in *Golden Terrace* (*Huang Jin Tai*) .❸

原文 Version of the Shamao with rounded wings, indicates dishonest characters.❹

拙译 有着圆形帽翅的纱帽，暗示了角色人物并非诚实可靠。

❶ BONDS ALEXANDRA B. Beijing Opera Costumes: the Visual Communication of Character and Culture [M]. Honolulu: University of Hawaii Press, 2008: 254.

❷ 阚艳华，董新颖，王娜，等. 中英文对照京剧服饰术语 [M]. 覃爱东，陶西雷，校译. 北京: 学苑出版社，2017: 163.

❸ 同❷163，略有改动.

❹ 同❶358.

原文 Yuansha are Shamao with rounded wings; these are worn by foolish, dishonest, and funny officials who may take bribes. They are played by Chou actors.❶

拙译 圆纱是指圆形展翅的纱帽，是昏庸糊涂、滑稽阴险的小人或贪官污吏的象征，属于丑角。

20. 尖展纱帽

原文 "尖展纱帽"又称"尖纱""奸纱"。展翅有菱形、桃形两种，展面黑漆底贴金或贴金点绢，镂古文、"寿"字等图案，主要为狡诈、贪婪、凶残的官员所戴。如《四进士》中的顾读、《一捧雪》中的严世蕃等。另外，具有鲁莽、豪爽性格的人物也戴此帽，如《牛皋下书》中的牛皋、《古城会》中的张飞等。❷

译文 Jianzhanshamao (Pointy-winged Gauze Cap) is also known as the "Jianchishamao" (Jianchi also means "pointy wings") or "Jiansha" (the "Treacherous Gauze [Cap]"), and is one kind gauze cap used by cunning, greedy and murderous officials. Gu Du in *Four Successful Candidates of the Imperial Examination* (*Si Jin Shi*) and Yan Shifan in *A Handful of Snow* (*Yi Peng Xue*) are two examples. Not only is it used by these vile characters, but it also worn by some roles with reckless but good-hearted character, such as Niu Gao in *Niu Gao Deliveries the War Declaration* (*Niu Gao Xia Shu*) and Zhang Fei in *Reunion at Gucheng Town* (*Gu Cheng Hui*). The wings of the cap come in two shapes: diamond and peach. The wings are gilded on a black background or decorated with golden decorations inlaid with satin.❷

原文 "sharp" or "pointed" Shamao-style hat; with peach- or diamond-shaped wings, indicating sinister or corrupt officials.❸

拙译 尖纱帽有桃形或钻石型的帽翅，暗示人物角色为贪婪或凶残的官员。

原文 Diamond or peach-shaped wings on the Jiansha (lit. "sharp or pointed" Shamao) indicate dubious characters, including traitors and corrupt officials in Jing and Chou roles.❶

拙译 奸纱上的菱形或桃形展翅（奸纱或尖纱）暗含人物角色可疑，常指扮演反贼和贪官的净、丑角色。

❶ BONDS ALEXANDRA B. Beijing Opera Costumes: the Visual Communication of Character and Culture［M］. Honolulu: University of Hawaii Press, 2008: 254.
❷ 阚艳华，董新颖，王娜，等. 中英文对照京剧服饰术语［M］. 覃爱东，陶西雷，校译. 北京: 学苑出版社，2017: 162.
❸ 同❶351.

译文剖析 纱帽为京剧中一般官员的头衣。中展纱帽、尖展纱帽与圆展纱帽都属于纱帽，不同纱帽帽翅的形状各不相同，代表不同角色人物。查阅《牛津高阶英汉双解词典》，"gauze"解释为"a type of light transparent cloth, usually made of cotton or silk，薄纱，纱罗（通常由棉或丝织成）"❶。译者将纱帽翻译为"Gauze Cap"；中展纱帽为京剧中俊扮的老生、小生角色或正派的官员选用，译文为"Medium Wing Gauze Cap"。在《牛津高阶英汉双解词典》中，"medium"的解释为"in the middle between two sizes, amounts, lengths, temperatures, etc.中等的；中号的"❷；"wing"则为"one of the parts of the body of a bird, insect or BAT that is uses for flying，（鸟、昆虫或蝙蝠）翅膀；翼"❸。尖展纱帽又称尖纱或奸纱，为狡诈、贪婪、凶残的官员所戴，其展翅呈桃形或菱形，译文为"Pointy-winged Gauze Cap"，其中"pointy"在《牛津高阶英汉双解词典》中的解释为"with a point at one end，尖的，由尖头的"❹。圆展纱帽帽翅呈圆形，取其圆滑之意，主要为昏庸糊涂的贪官或谄上骄下、吹牛拍马的势利小人以及奸诈狡猾、阴谋害人的人所戴。译文"Round-Winged Gauze Cap"中的"round"同鞑帽中的"round"，意思是"shaped like a circle or a ball，圆形的；环形的；球形的"❺。综上所述，三种帽饰的译文无论在外形，还是在材质等各方面，都在最大限度协调与原文的差异，通过阐释，凸显其人物特征，帮助读者理解其内涵，方便服饰文化的交流与传播。

针对美国学者的英文阐释，在汉译中采用文化协调，充分考虑其传递信息的着眼点与侧重点，句式结构符合汉语读者的阅读习惯，凝练简洁。

21. 渔婆罩

原文 "渔婆罩"又称"渔婆帽"。形似草帽圈，前檐上卷，帽檐外缘围以线网，四周缀丝穗或五彩珠串。前脸正中镶一大绒球，旁边缀小绒球。主要为渔女、村姑及其他劳动妇女所戴，如《小放牛》中的村姑。还有一种渔婆罩，又称"虞姬罩"。帽檐四周无线网，只垂彩穗。主要为宫内女辇夫、军营中的女旗手等人物所戴，如《霸王遇虞姬》中的虞姬。❻

译文 Yupozhao (Female Wide-rimmed Hat) is a decorative satin blue hat with a large pompon on the top, and embroidered with small flowers. The front brim is slightly curled, and adored with colourful strings of beads or tassels. It is worn by fisher women, young country women and other working women. For example, the village girl in the play *The Cowboy and*

---

❶ 霍恩比. 牛津高阶英汉双解词典［M］. 王玉章，赵翠莲，邹晓玲，等译. 7版. 北京：商务印书馆，2010：843.

❷ 同❶1254.

❸ 同❶2306.

❹ 同❶1528.

❺ 同❶1740.

❻ 阙艳华，董新颖，王娜，等. 中英文对照京剧服饰术语［M］. 覃爱东，陶西雷，校译. 北京：学苑出版社，2017：176.

the Girl (*Xiao Fang Niu*), and Yu Ji in the play *Xiang Yu Falls in Love with Yu Ji* (*Ba Wang Yu Yu Ji*) wear this hat. ❶

译文剖析 渔婆是指打鱼为生的劳动妇女，罩的本意是覆盖在表面罩住的意思。渔婆罩的字面意思为渔妇所戴的帽子，其译文为 "Female Wide-rimmed Hat"。其中 "female" 强调了头衣使用者的性别，并未表明属于渔婆专用，容易误导读者认为该帽属于普通妇女所戴，并非渔女、村姑及其他劳动妇女专属。"wide-rim" 则较好体现了渔婆罩宽檐的外形特征，协调了原文和译文在该头衣形制方面的差异，较好地再现了原文本意。

22. 大太监帽

原文 "大太监帽"，前扇扁圆，后扇为锥形。正中面牌用双龙戏珠图案，左右耳子垂挂流苏。有金、银两种，主要为太监（近侍）专用。如《法门寺》中的贾桂以及《贵妃醉酒》中的高力士、裴力士。❷

译文 Dataijianmao (Senior Eunuch Hat) is worn by the eunuch in charge. It is trimmed with the decoration of two dragons chasing a pearl on a Mianpai (a circular ornament used in costumes for acrobatic fighting roles), from which tassels dangle from both sides. The front section is shaped as an oval, and the back section is like a circular cone. It is made of gold-inflected or silver inflected silk. It is worn by roles such as Jia Gui in *Famen Temple* (*Fa Men Si*), and Gao Lishi in *The Drunken Concubine* (*Gui Fei Zui Jiu*) who wear gold hats with red tassels, and Pei Lishi in *The Drunkened Concubine* (*Gui Fei Zui Jiu*) who wears a silver hat with pink tassels. ❸

原文 Eunuchs have a distinct design for their court headdress that follows the style the eunuchs are wearing in a painting from the Ming dynasty in the National Palace Museum collection. The front of the hat fits the shape of the head, while the back has a piece that wraps around the back of the head and stands above the crown, curving into a cone shape. The Ming dynasty version is plain black, but the surface of the stage headdress is gold filigree, decorated with pearls and woolly balls. ❹

拙译 太监帽作为宫廷头衣，其设计独特，依照国家故宫博物院中明代的绘画藏

❶ 阙艳华，董新颖，王娜，等. 中英文对照京剧服饰术语 [M]. 覃爱东，陶西雷，校译. 北京: 学苑出版社，2017: 176，略有改动.

❷ 同❶177.

❸ 同❶177，略有改动.

❹ BONDS ALEXANDRA B. Beijing Opera Costumes: the Visual Communication of Character and Culture [M]. Honolulu: University of Hawaii Press，2008: 252.

品的样式。帽子的前部适合头形，后面则环绕头后部，弯曲成锥形且置于盔部上方。明代太监帽为纯黑色，但舞台上"头衣"的表面是金色掐丝的，并饰有珍珠和羊毛球。

23. 小太监帽

原文 "小太监帽"又称"老公卷"。样式同大太监帽，但尺寸较小。黑漆金边，装有绒球。四顶为一堂，主要为站班、摆驾或持仪仗的小太监使用，如《长生殿》中的四小太监。❶

译文 Xiaotaijianmao (Junior Eunuch Hat) is worn by eunuchs who carry an imperial flag and arrange the emperor's journeys.❷

译文剖析 汉语中的太监指古代被阉割生殖器后失去性能力，专供古代都城皇室役使的男性官员。在京剧中，根据品阶的高低，太监可被分为大太监和小太监，其头衣分别被称为大太监帽和小太监帽，对应的译文分别是"Senior Eunuch Hat"和"Junior Eunuch Hat"。查阅《牛津高阶英汉双解词典》，"senior"为"high in rank or status; higher in rank or status than others，级别（或）地位高的"❸；"junior"为"having a low rank in an organization or a profession，地位（或职位、级别）低下的"❹；"eunuch"为"a man who has been castrated, especially one who guarded women in some Asian countries in the past，阉人；太监；宦官"❺。原文中的大和小分别协调为太监资历的深浅和年龄的大小，暗示其身份等级的高低。太监协调翻译为"eunuch"，忠实于原文表达，表明人物形象，协调再现了原文内涵。

　　本章小结 | 本章基于京剧服饰中三衣，包括软片类（水衣、胖袄、护领、小袖、大袜、彩裤）、硬类（厚底靴、薄底靴、登云靴、皂鞋、彩鞋、鱼鳞鞋、打鞋、小孩鞋、僧鞋、花盆底鞋和福字履）以及头饰类（盔、冠、巾、帽），诸如梢子帽、毡帽、皂隶帽、风帽、毗卢帽、男僧帽、女僧帽、红缨帽、皇帽、鞑帽、侯帽、纱帽、中展纱帽、尖展纱帽、圆展纱帽、硬罗帽、鬃帽、大太监帽和小太监帽等诸多物品的英汉或汉英对照版本的阐释，充分挖掘其各自的文化内涵，结合文化语境，借助文化翻译协调原则给予协调表达，凸显作者和译者对于京剧服饰表达在"真、善、美"三方面的关照，力图推进不同时期、不同语境下的京剧服饰文化的对外传播。

---

❶ 阙艳华，董新颖，王娜，等. 中英文对照京剧服饰术语［M］. 覃爱东，陶西雷，校译. 北京：学苑出版社，2017：178.

❷ 同❶178，略有改动.

❸ 霍恩比. 牛津高阶英汉双解词典［M］. 王玉章，赵翠莲，邹晓玲，等译. 7版. 北京：商务印书馆，2010：1814.

❹ 同❸681.

❺ 同❸1101.

京剧服饰三衣类文化内涵及翻译协调美　第五章

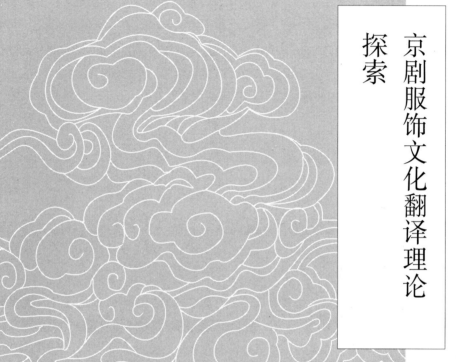

第六章

京剧服饰文化翻译理论探索

京剧艺术作为中华国粹，在长达一百七十余年的发展历程中，借助精美的服饰，演绎出数千部优秀的剧目，再现了中华传统文化的源远流长。面对当今文化语境，反思本书中（从第三章到第五章）有关京剧大衣箱、二衣箱和三衣类列举的京剧服饰术语英汉表达，如何正确对待京剧服饰文化翻译，在翻译理论的指导下实施京剧服饰文化翻译，并尝试京剧服饰文化翻译理论探索。

第一节
京剧服饰文化翻译反思

本书的第三章、第四章和第五章，全部聚焦京剧服饰术语。即使是同样的服饰，因为译者所处的时间段不同、身份不同、地域和政治历史背景不同，以及出版商不同、面对的读者群体也不同，译者各自的英文表达能力，包括对于京剧服饰文化内涵的理解也不尽相同，凡此种种，导致其英文表述各不相同。纵观本书中的四个研究文本，潘霞凤在1995年撰写出版 *The Stagecraft of Peking Opera*，易边（YI BIAN）在2007年出版 *Peking Opera: the Cream of Chinese Culture*，美国人亚历山德拉在2008年撰写出版 *Beijing Opera Costumes: the Visual Communication of Character and Culture*，以及2017年中央戏曲学院阚艳华、董新颖、王娜等撰写出版的《中英文对照京剧服饰术语》，在对比研究中发现，从1995~2008年，随着时间的推移，中华文化在对外交流的不断深入中，京剧服饰术语的英文表达日臻完善。2017年阚艳华等学者呈现的京剧术语汉英对照文本，甚至直接通过汉字加拼音给以阐释，在一定程度上更显"中国味"。需要指出的是：除去亚历山德拉是外籍人士，另外三个文本的作者都是华夏儿女，都在用各自不同的方式，助力承载着厚重文化的传统京剧服饰艺术"走出去"。那么，京剧服饰文化要"走出去"，究竟该怎样才能被外国读者或观众更乐于接受？

首先，要针对"走出去"的服饰文化内容进行甄别与选择，选择代表民族优秀文化基因，且能在当代西方观众心目中引起共鸣的京剧服饰进行文化协调英译。以位居京剧行头"榜首"的"富贵衣"为例，其主色调选用黑色，在上面点缀各种形状与颜色的小布块，使文化内涵与不同色彩完美融合。表面的补丁破烂呈现出错落有致的别样之美，内涵的阶段性贫穷与最终的翻身富贵，恰在矛盾中协调统一，在苦难与挣扎中隐含着坚韧与甜蜜，表达了主人公敢于面对困难、坚持不懈、改天换地的决心与信心。这与西方电影《当幸福来敲门》之类的励志故事有着异曲同工之妙，其主题思想容易为西方观众所理解接受。

其次，要摆正文化"走出去"的心态。译者要有豁达的心胸和气度对待文化差异，坚持"求同存异"的文化心理，尊重异域文化，正确客观看待本民族文化传统，既不要趾高气扬，也不要低声下气；而要不卑不亢，能够主动换位思考，以平和的心态看待自身的民族文化与世界上其他文化之间的差异。同样是服饰术语"富贵衣"，在西方世界的某些地域，可能人们仅局限于色彩的理解，也有可能只关注行头名称的别样，或许只想明白谁是着装人物、甚至衣服上破洞的形状与大小。但是国外的京剧爱好者也许还想知道"富贵衣"在汉语里怎么读、怎么写，凡此种种。所以我们只好"以不变应万变"，怀揣至诚之心，协调文化差异，促进文化真正地"走出去"。阙艳华等学者倡导的类似Fuguiyi（富贵衣，"wealth and honour clothing"）的英文表达，弥补了前三种译文中仅有的音译表达，汉字后面的阐释性翻译可以方便西方读者理解其文化内涵。但是如此英文表达的术语是否就是最佳英译，依然需要动态看待。或许有人认为，"clothing indicating the inspiration of prosperity"更合适。不同时代、不同语境、不同接受对象、不同出版商……太多的因素在决定或左右我们对于译文的选择。事实上，对于术语"富贵衣"，还有其他的诸如"海青"，又名"道袍"的俗称，甚至还可直接称为"穷衣"，不妨将其对应的英文阐释为"the monk's robe, Taoist robe, or the Qiong yi（the garment of the poor）"。但是，在前面讨论的四个英文表述中，亚历山德拉将其阐释为"the garment of wealth and nobility"，而阙艳华等学者则解释为"wealth and honour clothing"。查阅资料，nobility具有贵族和高贵品质的含义，honour更加强调荣誉；而garment和clothing比较，后者则强调某类特色服装。笔者为此请教远嫁美国二十余年的朋友、学者，探讨该术语的英译，认为"beggar's gown"合适，因为gown本身就强调了着装者的身份和地位，长袍是有地位的象征。如果需要在nobility和honour之间选择，前者更为贴切，而且似乎将亚历山德拉和阙艳华等学者的意见综合为"the clothing of wealth and nobility"更为妥帖。孰是孰非，我们暂且不加评论，但是作为京剧服饰术语，那就意味着标准规范和名称统一。如果不是其原来的英文表达谬以千里，我们不妨依照时间顺序暂且"长者为大"。因为"多名并存"的英译只会使西方读者一头雾水，在一定程度上阻碍文化"走出去"。类似的文化术语表达，如"荷包"的英译表达，笔者就曾综合其诸多功用，查阅杨宪益夫妇和大卫·霍克斯（David Hawkes）在《红楼梦》译本中不同语境下的差异化表达，在《外国语文》撰文呼吁术语的规范统一、音译并采用括号加注，不失为最佳选择之一。

最后，在文化"走出去"中要注意把握节奏，注重传统文化与时尚文化的融合，积极主动寻求民族文化与世界文化元素共同的对接点。古老的京剧服饰文化艺术，跨越历史，历经风雨，传承至今，走向未来，如果操之过急，一味追求原汁原味，保持原生态，则可能在一定程度上不容易得到异域读者的理解与认同。这就需要我们在英文表述中结合当代文化元素，针对传统元素给予不同程度的协调处理，包括当今文化语境下通过强调富贵衣补丁的不同色彩（红、绿、黄、粉色等）与形状（三尖块、方块、圆块或是葫芦形块），暗喻多元文化协调之美。当年的国家领导人曾将传统剧目《梁山伯与祝英台》直接解释为类似西方名剧《罗密欧与朱丽叶》，方便外宾理解其故事梗概。同样，今天我们也可以采用文化类比或文化替代的翻译协调策略，在译文中恰当借用一些读者耳熟能详的词句来表达原文的某些文化词汇或概念，将京剧《珍珠衫》解释为西方观众熟悉的莎剧中"致命的手绢"——《奥赛罗》。同时补充说明两者的结局完全不同，因为我们传统剧目的结局大多是大团圆，弘扬人间的公平、正义和"真、善、美"，鼓励人们坚信善有善报；而《奥赛罗》则是典型的爱情悲剧，结局凄惨，令观众唏嘘不已。但是面对新时代的观众，故事的本身以及唯美的服饰尚能吸引部分青年观众，而情节的发展则会引发类似梁山伯十里相送，为什么不表白？因为当代青年认为"爱就要大声说出来"。面对诸如此类的发问，在文化"走出去"的过程中，我们只能尝试在翻译中采用协调的手段来增补原文所承载的额外或附带信息，针对类似京剧术语"富贵衣"的英文介绍给以符合时代要求的全面说明，动态协调文化差异，逐步帮助读者理解接受。京剧服饰文化术语蕴藏着深厚的文化内涵，启人深思。独具特色的翻译实践的探索需要在文化翻译理论的指导下不断探索，以成就和谐译文。

第二节
文化翻译理论探索

语言是文化的载体。语言和文化之间相互依存，语言翻译既是两种语言的转换，也是两种文化信息之间的转换。其核心问题在于：一是怎样处理好民族传统文化，二是怎样处理好外国文化。民族传统文化需要传承与创新，而外国文化则需要尊重与扬弃。源语文化与译语文化实际是处于融合共生、共同发展繁荣的态势。

（一）文化翻译研究基础

王佐良先生曾强调译者必须是一个真正意义的文化人。从跨文化交流的角度反观翻译，其本质就是文化再现，再现源语文化特色。鲁迅先生曾经指出翻译必须保存着原作的丰姿，对于源语文化，不可以任意抹杀和损害，译者在翻译时不要拘泥于原文的字面意思，要深刻理解原文所承载的文化信息，力求保持源语文化的完整性和一致性，并在译文中再现出来。在具体翻译过程中，要坚持文化信息的真实性，实现对源语文化益智成分的保留和阻抗式的翻译策略，以保留和反映外国民族语言与文化的特点，使目标语读者理解传统的民族语言和文化的差异，有效促进文化交流，丰富本土文化。翻译的目的不仅在于求同，也在于存异。译文的价值取决于其语言差异的反应程度以及对该差异的强调程度。

在综合目前国内学者研究的基础上，笔者尝试将其文化翻译归纳为三类：

一是对翻译本质的认识。苏联翻译家巴尔胡达罗夫（Barhudalov）在《语言与翻译》一书中曾说过，翻译一词用于从一种语言翻译成另一种语言时有两层意思，一是指"一定过程的结果"，二是指"翻译过程本身"，即翻译这一动词表示的行为，而这一行为的结果则是译文❶。因此，翻译在基于理解和转换的基础上，译者需要对于获得的审美感受与认识做细化整理。何刚强认为，中文的翻译是指"翻"和"译"的融合，"翻"的本意是挪动或折腾，"译"的本意是解释和说明❷；英文的"翻译"表达则至少可以用七八个不同的单词，比如"transfer""transform""transmigrate"或"transfuse"等，暗含翻译需要根据具体情况，进行变动、调整和变通，翻译的本质是沟通。翻译是文化介入与文化协调。❸

二是对翻译目的的认识。克罗宁（Cronin）认为，在翻译的三角关系中，处于第三方的翻译或译者帮助A语和B语的使用者实现交流，开启彼此之间的对话，确保各方权利、名誉或身份不受威胁。在种族对立或冲突中，蕴藏在翻译中的斡旋有助于消除潜在的暴力。对任何一方来说，真正的挑战在于通过翻译来维护差异，而非通过暴力来消除它❹。

三是对翻译策略选择的认识。王沪宁认为，文化是一个国家软实力的重要组成，软实力的力量来自其扩散性，只有当一种文化广泛传播，软实力才会产生强大力量❺。正如加纳尔（Ganahl）所言，英语已经成为"一种过渡语，一种移动的语言"❻，翻译策

---

❶ 巴尔胡达罗夫. 语言与翻译［M］. 蔡毅，虞杰，段京华，编译. 北京：中国对外翻译出版公司，1985：5.

❷ 孙艺风，何刚强，徐志啸. 翻译研究三人谈(上)［J］. 上海翻译，2014(1)：10.

❸ 廖七一. 翻译研究：从文本、语境到文化建构［M］. 上海：复旦大学出版社，2014：66.

❹ 张慧琴，武俊敏. 服饰礼仪文化翻译"评头"之后的"论足"——基于《金瓶梅》两个全译本中的"足衣"文化［J］. 上海翻译，2016(06)：54.

❺ 王沪宁. 作为国家实力的文化：软实力［J］. 复旦学报(社会科学报)，1993(3)：91—96.

❻ GANAHL R. Free Markets：Language，Commodification，and Art［J］// E APTER. Translation in a Global Market，Special issue of Public Culture，2001(1)：29.

略的选择要遵循中山（Nakayama）和哈卢拉尼（Halualani）提出的平等三原则，即语言平等原则（避免语言歧视）、文化平等原则（任何人都有权使用自己的语言和文化）和信息交流平等原则（任何人都有权表达观点和信息）❶，在跨文化交际中应相互尊重对方的语言和文化，并注重保留交际者本族文化的特色❷。

上述三个层面关于文化翻译的认识与实践，是文化翻译协调的核心，其关键在于以介绍中华文化为手段❸，深刻认识自己的文化，理解并接受多种文化，在多元文化的世界格局中确立自己的位置。同时，"在不损害中国文化精神的前提下，探索最合适的方式来解读和翻译最合适的典籍材料，从而达到消解分歧，促进中外文化的交流，极大地满足西方受众阅读中国典籍的需要"❹。这种文化的自觉与自省，其最终目的是促进自我了解与自我完善❺。

（二）文化翻译原则与方法

任何翻译都因文化差异而存在，因而在一定程度上又都属于是文化翻译。从实践层面看，文化翻译是指具体文化信息的翻译；从理论层面而言，文化翻译研究属于在宏观上讨论翻译中的文化策略，在微观上讨论翻译中文化信息的处理。任何一种文化翻译理论都起源于民族文化，尤其是文化翻译思想的传承以及个人对文化翻译在宏观和微观上的思考。文化翻译理论既是基于文化角度对翻译的考察，也是源自翻译角度对文化翻译理论的思考。前者着重考察翻译在文化中的地位（考察的对象是翻译），后者强调文化对翻译的影响（考察的对象是文化）。

从历史的角度而言，文化翻译理论具有流变性与传承性❻。流变性表现在文化翻译理论的不断发展，传承性则体现在任何文化翻译理论都有其历史渊源。从共时的角度而言，由于各种语言、文化都有时代性和区域性，因而任何一种文化翻译理论往往都具有时效性和区域性。不同文化翻译理论之间既有共性又有个性。就共性而言，它们都聚焦跨文化翻译的问题，同时也都具有民族性和兼容性。文化具有民族性，文化翻译理论同样也具有民族性。文化是兼容的❼，不同民族的文化翻译理论同样相互兼容。不同文化翻译理论的个性同样体现在其民族性之中。民族主义必然或强或弱地存在于各种文化翻译理论中，任何文化翻译的指导原则都不会背离本民族的利益。

❶ NAKAYAMA T K, HALUALANI R T. The Handbook of Critical Intercultural Communication [M]. Chichester, West Sussex: Wiley-Blackwell, 2010: 248-270.

❷ 郑德虎. 中国文化走出去与文化负载词的翻译 [J]. 上海翻译, 2016(2): 53-54.

❸ 黄友义. 中国特色中译外及其面临的挑战与对策建议——在第二届中译外高层论坛上的主旨发言 [J]. 中国翻译, 2011, 32(6): 5-6.

❹ 罗选民, 杨文地. 文化自觉与典籍英译 [J]. 外语与外语教学, 2012(5): 63-66.

❺ 张南峰. 文化输出与文化自省——从中国文学外推工作说起 [J]. 中国翻译, 2015(4): 93.

❻ 刘宓庆. 口笔译理论研究 [M]. 北京: 中国对外翻译出版公司, 2004: 13-20.

❼ 同❻20.

在具体翻译实践中，刘宓庆认为，文化翻译应遵循以下三个原则：（1）文化信息的表现应该适应目的语文化现实和发展之所需，即文化适应性原则。（2）文化翻译的基本要义应坚持实事求是，即文化科学性原则。（3）关涉翻译成败的，在文化表现中对于原文有着至关重要的审美判断力，即审美判断原则。❶文化翻译的任务在于翻译容载或含蕴着文化信息的意义，文化翻译研究必须围绕意义问题展开。语义诠释任务的圆满解决，取决于译者对文化内涵的多维观照，主要有以下四个方面：文化历史观照、文本内（上下文）证与文本外证、互文观照、人文互证（即根据作家本人来求证）。语义文化论证的原则是以证定义，其中包括求证、参证（即避免传讹）与举证。文化翻译理论的核心课题是：语义的文化诠释、文本的文化解读、文化翻译表现论、翻译与文化心理探索。文化翻译理论中最富于铺垫性的深层课题是语言文化心理问题。

文化翻译的方法论体现在对于文化信息的给予，进行图像、模仿、替代、阐释和淡化等方式的翻译处理。表现法的取舍通常必须进行语义考量（看哪一个对应式更切合原意）、语用考量（交流中合不合"用"）、审美考量（实质上也是一种语用考量，着眼于审美效果）、政治考量（也属于语用考量，实际上是一个文化适应性问题）。最佳可读性翻译通常是综合考量的结果。做到"原汁原味"是文化翻译的表现论思想，其实质体现在译者必须精于把握原语文本的特征和文化气质，模仿原语文本的结构和原文特征，并精于做到由意义、意向来决定，而意义、意向又由语境中的"用"来调节、变通，以保证原语文化对译语文化的适应性。或者说，译者必须提高译语的表现力以适应或吸收原语语言和文化的活力。总之，译者必须精于对原语文本的动态模仿。翻译要确保"原汁原味"就要从微观和宏观上准确把握作品的个体文化特色和群体文化特色。文化翻译的表现法是一个开放系统，不存在任何公式化范式。

针对文化翻译的效果论（目的论）和价值论，刘宓庆认为，文化翻译应该定位于文化战略手段，服务于国家或民族的核心利益❶。文化翻译是一种文化互动而不是简单的同化，是不同文化之间相遇、相识和相知的过程。当两种或多种文化相遇时，有交流、汇合、融合、分解、斗争、抗拒、接受、拒绝，形态各异。一国文化传入另一国后，往往要经历一个适应过程。适应就意味着要改变，依据新环境进行调整以适应当地之需。❷孟华认为，任何一种相异性，在被植入一种文化时，都要做相应的本土改造❸。找到与原文对应，又能为本民族读者理解或接受的词语来进行置换。站在文化的角度，翻译在一定程度上都是文化协调的过程与结果。正如刘宓庆、章艳所述，文化翻译比较复杂，在理论原则上只能采取描写主义态度，按照语言现实和文化审美现实，在纷繁的语言文化现象和翻译事实中寻找带有规律性的参照规范而不是原则，任

❶ 刘宓庆. 文化翻译论纲(修订本)［M］. 北京：中国对外翻译出版公司，2006：245.
❷ 季羡林. 季羡林论中印文化交流［M］. 北京：新世界出版社，2006：27-28.
❸ 孟华. 翻译中的"相异性"与"形似性"之辩——对翻译与文化交流关系的思考与再思考［C］//谢天振. 翻译的理论建构与文化透视. 上海：上海外语教育出版社，2000：199.

何"非此即彼"的规定都无济于事❶。源语和译语之间的文化经验相同或相似的可能性越大，彼此契合或对应的可能性也就越大，反之亦然。文化词语流行的程度（使用频度和流通版图）越大，模仿或直译的条件也就越充分。

无独有偶，美国学者安德烈·勒菲弗尔（Andre Lefevere）认为："翻译的首要任务是要使译文读者看得懂。理想的翻译既能表达原文的意义，又有能表达原文言外意义的力度。但在实践中，一般往往只能完成第一个任务。"❷因此，在翻译中要承认不可译的存在，但不能轻易妄下不可译的结论，而应坚持"适译"文化翻译观，协调处理文化差异。"适译"文化生态翻译观倡导"人类厚生文化万事"❸，强调"同则不继"与"生万物"，文化之间如果只是在"近亲"（同种、同族、同类）之间交流，则有退化与灭种的危险；相反，适当比例异质因子的"和合"则有利于物种的进化。"适译"文化生态翻译观的本质是"适量""适宜"与"适度"的和谐统一，是译文和原文彼此跨越文化差异、实现诸多矛盾和谐统一的结局。这其中最难的是对"度"的把握。❹"适量"翻译要从本土文化生态系统的稳定性与和谐性出发，在通过翻译进行的对外文化活动中做到"适量吸收、以我为主"，捍卫民族文化的主权与完整，保证本土文化的创新，抵制与反对强势文化的侵略与殖民；"适宜"翻译要有本着保护目的语文化生态系统发展与创新的文化因子的原则，避免引进易导致目的语文化生态系统产生极大混乱或巨大破坏的内容，杜绝引进的内容可能会直接替代目的语文化生态的文类系统或价值体系，尤其是核心的文类与价值；"适度"翻译要对适宜的文化适量引进，使适宜翻译和适量翻译协调统一。❺这就意味着译者在翻译过程中需当好"读者"——准确地理解原文；当好"判官"——得体地衡量翻译生态环境；当好"伙伴"——合作地兼顾"诸者"（如作者、读者、出版者、资助者、委托者、译评者等）。❻同时，译者在文化立场上应明确自己的国家身份、民族身份和地域身份，在翻译策略上必须把对语言的字面转换拓展为对本土文化内涵的阐释，促进本民族文化不断更新、与时俱进。❼在"物相杂而生生不息"理念的指导下，实现和谐、多元与创生的文化生态理想。

翻译作为一种社会现象，孕育于社会环境中，译作的产生、接受与传播必然受当时社会环境中诸多因素的制约。这些因素相互联系、斡旋互动，彰显各自在翻译过程中的功能，使翻译成为"受社会调节之活动"❽。同时，"翻译是一个自我适应性强，自我调整和自我反省，并能自我再生的系统"❾。作为社会系统的子系统，"深嵌于社会、

❶ 刘宓庆，章艳. 翻译美学理论［M］. 北京：外语教学与研究出版社，2011：146-152.
❷ 陈光祥. 可译性与可译度［J］. 外语研究，2003(3)：57-60.
❸ 张立文. 和合学［M］. 北京：中国人民大学出版社，2006：635.
❹ 张慧琴. 高健翻译协调理论研究［D］. 上海：上海外国语大学，2009：23.
❺ 吴志杰. 和合翻译研究刍议［J］. 中国翻译，2011（4）：5-13，96，略有改动.
❻ 胡庚申. 翻译适应选择论［M］. 武汉：湖北教育出版社，2004：123.
❼ 张景华. 全球化语境下的译者文化身份与汉英翻译［J］. 四川外语学院学报，2003(4)：128-130，略有改动.
❽ THEO HERMANS. The Conference of the Tongues［M］. Manchester：St. Jerome Publishing，2007：10.
❾ THEO HERMANS. Translation in Systems：Descriptive and System-oriented Approaches Explained［M］. Manchester：St. Jerome Publishing，1999：141-142.

政治、经济和文化环境里，构成整个系统内部的一种操作力量"❶。任何翻译在一定程度上都是文化协调的翻译。事实上，早在20世纪80年代，西方学者霍尔茨·曼塔利（Holz Manttari）就以"跨文化合作"（Intercultural Cooperation）来看待翻译；勒弗维尔（Lefevere）把翻译看作是"文化交融"（Acculturation）；而斯内尔·霍恩比（Snell Hornby）则认为翻译是一种"跨文化的活动"，并建议从事翻译理论研究的学者抛弃"唯科学主义"态度，把文化而不是文本作为翻译单位，把文化纳入翻译理论研究中。❷我国学者杨仕章则认为翻译是一种跨文化交流；翻译的目的是突破语言障碍，实现并促进文化交流；翻译的实质是跨文化信息传递，是译者用译语重现原作的文化活动；翻译的主旨是文化移植与文化交融。❸陈历明也同样认为，翻译看似只是从事语言文字的转换，实则总是要跨越时空，在两种文化的夹缝中穿行游走，为两种文化的和谐对话，不断协调，牵线搭桥。❹作为译者，在原作与读者之间来回调解、平衡，决定对话中的话轮趋向，其承担的角色不仅是两种语言的内容、形式、风格的调解者，更是两种处于不同时空的文化对话、交锋、融合、斗争与妥协的调解者和协调者。究其实质，正如孙艺风所述，文化翻译是一种文化互动而不是简单的同化❺。翻译的衍生性和调节作用意味着跨文化翻译是阐释的具体化，而不是文化形式的直接转换，是文化协调的结果与结局。人类文化的千差万别使历史上没有哪种文化能仅靠自身的力量生存、发展、壮大。文化的对话源于彼此的需要，各文化间共性的存在，即文化"共核"（Common Core）是文化对话的基础，也是文化协调翻译进行的前提。译者正是通过对整个社会环境中各种矛盾的文化协调，使社会环境中的不同文化在千差万别中和谐并存。

第三节
京剧服饰文化翻译理论探索

（一）京剧服饰文化协调翻译的理据

回顾历史，我国译者本身就具有天然的协调意识。中国哲学讲"天人合一"，而"中和"观念是人们对自然界万事万物之间复杂关系的审美、认知方式。《礼记·中庸》

---

❶ 李红满. 布迪厄与翻译社会学的理论建构［J］. 中国翻译，2007(5)：6–9.
❷ 张慧琴，徐珺.《红楼梦》服饰文化英译策略探索［J］. 中国翻译，2014，35（02）：111.
❸ 杨仕章. 文化翻译观：翻译诸悖论之统一［J］. 外语学刊，2000(4)：66–70.
❹ 张慧琴，徐珺.《红楼梦》服饰文化英译策略探索［J］. 中国翻译，2014，35(02)：111.
❺ 孙艺风. 翻译与跨文化交际策略［J］. 中国翻译，2012，33(1)：20.

推崇的为人之道是："喜怒哀乐之未发，谓之中，发而皆中节，谓之和。中也者，天下之大本也；和也者，天下之达道也。致中和，天地位焉，万物育焉。"❶基于这种哲学思想，协调作为翻译观念，它是一种意识；作为一种艺术法则，它是一种规律；从方法论的角度看，也可以说它是一种方法。正如"协"字本身所承载的文化内涵，"协"从"心"、从"力"，同心才能协力。这就要求译者不能亦步亦趋地紧跟原作，而是应该协调原作中各种因素和各方面的关系，在各种差异与对立的因素中寻求互相呼应与协调统一。这就要求译者在翻译实践中必须从所面对的双语出发，对双方各自的语性予以承认、理解、重视、服从、执行、体现、贯彻，从而使所译的原语一方的种种特点能够在译语一方的有关条件允许的情况下，通过必要和积极的协调，以一定程度改变了的形式重新在归宿语中得到再现，以便在总的精神和大体效果上取得与其出发语比较一致、基本吻合的效果。

德国功能学派理论家诺德（Nord）曾指出："译者在具体的翻译过程中既要忠诚于译语读者和翻译发起人，又要忠诚于原文作者，以达到两者之间的平衡，协调译入语文化和译出语文化对翻译的制约作用。"❷其实质是在翻译发起人和译文接受者之间发挥协调作用❸。同样，哈贝马斯（Habermas）在交往伦理中提出的协调性规则也同样体现出翻译协调思想。他认为，协调性规则作为人们在长期社会实践中逐渐形成的社会规范，带有契约性质，人们只有遵守这些规范与规则，才能在平等交流、相互尊重、相互理解的基础上顺利进行交往。这与翻译中原作与译作只有在相互协商、协调处理译文与原文中的若干矛盾的基础上，才能在译文与原文中尽力达成和谐的原则有着诸多的不谋而合。❹

究其实质，翻译是一种跨文化的交际活动，也是一种跨文化的对话活动。但是由于两种文化的不同，在对话活动中有时也会发生碰撞与冲突。任何目的语文化都倾向于积极接受和吸收与自己的"恰恰调和"。译语文化的影响是一种传统势力，它像一只强有力的大手左右着每一个译者的审美意识和表达方式。译者想要摆脱传统观念，就需要理论上的照亮和创新意识的指引。国内外翻译理论的发展，迫使我们改变以往的传统观念。钱钟书先生曾经指出："传统有惰性，不肯变，而事物的演化又迫使它以变应变，于是产生了一个相辅相成的现象。传统不肯变，因此惰性形成习惯，习惯升为规律，把常然作为当然和必然。传统不得不变，因此规律、习惯不断地相继破例，实际上做出种种妥协，来迁就演变的事物。"❺译语的异化是异国语言和文化对汉语的侵蚀，我国固有的语文习惯在外国语习惯的冲击下不断地相继破例，做出妥协。所以，一开始译语与汉语语文习惯的某些差异可能会给读者的感官造成审美障碍。这种障碍

❶ 徐行言. 中西文化比较［M］. 北京：北京大学出版社，2004：330.
❷ NORD CHRISTIANE. Text Analysis in Translation［M］. Amsterdam New York：Rodopi Bv Editions，1991：126.
❸ 张沉香. 目的理论与应用翻译研究［M］. 长沙：湖南师范大学出版社，2008：124.
❹ 吕俊，侯向群. 翻译学：一个建构主义的视角［M］. 上海：上海外语教育出版社，2006：146.
❺ 钱钟书. 中国诗与中国画［J］. 中国社会科学院研究生院学报，1985(01)：1.

久而久之又被人们所习惯，变成新的审美趋势。人们在阅读国外的戏曲翻译作品时，那种由译语的异化造成的情调或许给人以新奇的美感。因此，翻译传统京剧服饰文化艺术的过程中，要根据本土文化的需要，充分利用语言文化的开放性与渗透性、包容性与容纳性，展现原文特有的文化思想和艺术特色。通过不断协调英汉语言的文化差异，促使原语文化与译语文化之间互涉、互证、互补与互融。

究其本质，京剧服饰文化翻译的实践过程是一种寻求主客观统一的过程，一定的社会生活的映象和服饰文化内涵都客观地存在于原作之中。但是，这种客观的东西只有经过译者头脑的加工制作，以及经过译者主观能动性的创造作用，才能在译文中将其重新客观化和物质化。从这个意义上来说，服饰文化翻译不但是原作者对于服饰文化内涵的理解表达，更是译者对于服饰文化内涵在异域文化土壤中赋予新生命的产物。因此，翻译家的主观能动性不但是不可避免的，而且是不可缺少的。一切企图排除译者主观能动性的观点，都是形而上学的观点。因为从形而上学的观点来看，译者世界观和作者世界观的矛盾，似乎永远无法化解。但是，从辩证法的观点来看，这一矛盾却又是可以解决的。因为，如果从历时性（纵向）上视之，文化就是人类或特定种族的生命，它如同单个个体的生命一样，是一个不断发展、生生不息、连续不断的绵延过程，他所展现的是整个人类或种族的生命活动。❶

事实上，只有具有先进世界观的译者，才能把原作者按照他自己的世界观所反映的社会生活的本质方面和规律，真正客观而充分地揭示出来；只有具有先进世界观的译者，才能把原作的思想和艺术价值，真正客观而充分地挖掘出来。龚光明针对文化误读的成因有过精辟的论述，"人们与他种文化接触时，很难摆脱自身的文化传统和思维方式，往往只能按照自己熟悉的一切来理解别人。人在理解他文化时，首先按照自己习惯的思维模式来对之加以选择、切割，然后是解读。这就产生了难以避免的文化之间的误读"❷。翻译艺术中存在一个奇妙的辩证法，并为无数翻译实践所证实，接近原著有时反而脱离原著，脱离原著有时却是接近原著。❸这就意味着，在服饰文化翻译中，译者和作者的世界观要和谐一致，要忠实原文，体谅原作和译作读者之间的文化差异，应满怀善意地对待文化差异，跨越时空界限来解读文化差异，不断协调文化差异。

"京剧服饰"作为传统服饰文化翻译中的重要组成，在当今文化语境下的服饰文化翻译，不仅要对中国传统京剧服饰文化有正确的理解，而且要使文化的交流着眼于能真正影响普通人精神生活的层面。在文化语境下的服饰翻译中，一方面要注重文化差异，同时也要注重读者的感受。正如郭建中所说："跨文化交际的目的之一是增进不同文化之间的了解，这一目的自然也要求我们在翻译中要注意反映民族文化所特有的规范或风情。仅仅追求内容上的同一性，而无视蕴含在不同形式中的美学内涵和文化差

❶ 李炳全. 文化心理学［M］. 上海：上海教育出版社，2007:114.

❷ 龚光明. 翻译思维学［M］. 上海：上海社会科学出版社，2005：138.

❸ 刘文捷. 翻译入门：英译汉［M］. 成都：西南交通大学出版社有限公司，2004：10.

京剧服饰文化翻译理论探索　第六章

异，一方面会抹杀原作者的艺术创造，另一方面也会模糊不同文化之间的差异，造成不必要的文化误解。"❶ 其翻译过程中首先应以平和的心态，坚持文化相对主义的立场，注重文化概念的译介。要适时、适度、适量地结合文化语境，不断进行文化协调，❷ 科学把握文化语境的解构或重构。既不要根据自己的理解，过分添加不必要的冗长解释；也不要因为想当然地认为读者理解原文文化而忽略对其丰富内涵的阐释。而应尝试着以原文的语境为横坐标，读者的阅读语境为纵坐标，探索横纵坐标的"交汇点"，进行文化协调。基于语境，动态处理文化特色词汇，注重体现文化内涵，以实现最大程度的文化"保真"。❸

（二）京剧服饰文化协调翻译内涵

京剧服饰文化翻译极其复杂，翻译文本的选择，文化翻译过程的实施，以及翻译的结果，实际上都是协调的产物。协调是一种寻求译者与作者、作者与读者、译者与读者双方或多方都能接受的妥协方案和折中办法。❹ 在翻译实践中，当两种语言进行对译时所产生的译语，充其量也只可能是这两种语言的一种调和物，因而既不能过于依从原语的表达（这样就成了翻译腔），也不能本国化得太厉害（这样又成了归化翻译）。而只能是既很像源语，又读起来大有本国语的味道，即协调的产物。所以翻译不仅是反映，不仅是传达，不仅是照搬，不仅是释意、模仿、改写……而首先是一定矛盾的协调，翻译即是协调。虽然人类的思维方式是相通的，语言具有共性，但是不同民族语言之间的差异毕竟是客观存在的。奈达（Nida）曾说过："某一语言在特定领域内有大量相应的词汇，它们是文化焦点的重要标志。例如，大多数西欧语言有比例极高的技术词汇；苏丹的安努亚克人的语言中有好几百个词语是用来描述不同种类和特征的牛的；秘鲁高原的盖丘亚人对于不同种类和形状的马铃薯就有十几个不同的名称。"❺ 即使是我国大陆和台湾，在很多术语的表达方式方面也存在着明显的差异。譬如，橙子（大陆）和柳丁（台湾）、章鱼（大陆）和花枝（台湾）、软件（大陆）和软体（台湾）、鼠标（大陆）和滑鼠（台湾）等。在翻译过程中，具体的字斟句酌要因译文读者的差异进行协调，可谓"见大陆人说大陆话，见台湾人说台湾话"。

京剧服饰文化翻译中要以理解原文与原文作者、目的语读者及其所处的社会文化背景为前提，绝非简单意义上的"看懂"或没有生僻字，而是涵盖多个层级和维度的思维活动，跨越语言文字、民俗传统、宗教信仰、伦理道德和价值观念等差异，实现译者对于原语的多维解释，潜含着意义、审美和逻辑等多个视角。在理解阶段，首先

❶ 郭建中. 翻译中的文化因素：异化与归化 [M]. 北京：中国对外翻译出版公司，2000：207.

❷ 张慧琴，徐珺. 服饰文化翻译研究从"头"谈起 [J]. 中国翻译，2012(3)：109—112.

❸ 张慧琴，陆洁，徐珺.《红楼梦》中"荷包"文化协调英译探索 [J]. 外国语文，2013(3)：144.

❹ 张慧琴. 高健翻译协调理论研究 [D]. 上海：上海外国语大学，2009：16—17.

❺ NIDA E A. Language，Culture，and Translating [M]. Shanghai：Shanghai Foreign Language Education Press，2001：105.

要把握交流中的语境，成就交流中的意义，对原文形成正确且准确的认识；同时，结合语境，使原文的意象与情思实现艺术情景化，协调译文和原文之间存在的矛盾，力争使原文中隐约若现的含义实现语义的全面整合，并考虑到原文的美学韵味，关照到译文读者的接受能力，重视文本翻译后所达到的交际效果和目标。从理论角度而言，可以概括为对原文文本从审美角度给予认识，逐步融合为理解的感知、想象，并融入情感，逐层推进和深入，上升到文化层面，实现从表面的感知到历时和共时的分析，以殚思竭虑的精神和一丝不苟的态度吃透原文。正如刘宓庆所言，理解是对翻译全过程之发展起推动作用，使之具有转化的可能与潜势和动势。❶

在具体的京剧服饰文化翻译中涉及的问题既包括原作中文化信息的理解与传达，也包括民族文化对译者翻译策略的影响和译作的双文化的选择。原作中特定的服饰文化，暗含着该民族的文化内涵。想要准确地把握和理解原作京剧，不仅要理解原作服饰文化所包含的各种文化信息，还需要借助原作之外的文化背景知识。"要想进一步阐释艺术文本，必须考虑到民族心理与有声言语行为（以及无声言语行为），因为这些成分在很大程度上决定了社会心理活动的特性与本质。"❷对原作中的服饰文化因素和服饰文化信息归类剖析，根据其跨文化交际活动中的交际价值来确定是否翻译和如何翻译等。译者在服饰文化因素的处理过程中必然会受到译语文化的影响，总会自觉或不自觉地考虑到译语文化的特点，根据翻译所要实现的交际目采取相应的翻译策略。译者的翻译策略不仅体现在文化因素的翻译方面，还表现在译者对原作的整体文化态度上。这是源语文化与译语文化两者共同作用的过程与结果。

服饰语言本身具有民族文化内涵，服饰语言的使用也具有民族特色，反映出不同地域和民族的风俗、制度、道德、心理等不同方面的内容。因此，京剧服饰文化翻译牵涉到文化学、跨文化交际学和文化语言学，只有把原作和译作分别放到源语文化和译语文化两个不同的文化背景下，将翻译看作是一种文化的接触、交流与适应，才能认识翻译所含的文化学意义以及它在发展过程中所受到的各种社会文化的影响。同时，翻译活动是一项由人参与的文化活动和跨文化交际活动，译者作为该活动的主体，其所具有的文化身份以及对于文化因素的理解都对翻译产生影响。正如 S．西蒙（Simon）所指出的那样，"翻译是一场'战役'，既要最大限度地保障交际的通畅，又要克服语言上的社会文化差异，并且不要抹杀原文的语言文化特色"。❸如何在克服文化差异的同时，又保留文化特色是翻译学中最具争议的问题，同时也是实践中最难完成的一个任务。❹翻译是将处于源语文化中的原作首先转换成译语，再将译语移植到译语文化当中的一个复杂过程。在这一过程中首先要准确理解原作中所包含的各种文化信息，其次

❶ 刘宓庆．新编当代翻译理论［M］．北京：中国对外翻译出版公司，2005：72-79.
❷ 杨仕章．文化翻译观：翻译诸悖论之统一［J］．外语学刊，2000(4)：66-70.
❸ 张慧琴．高健翻译协调理论研究［D］．上海：上海外国语大学，2009：17.
❹ 何晓婷．试论文化翻译理论在高校日语翻译教学中的运用［J］．科技视界，2013(06)：96.

要根据译语文化处理其中的文化内涵。

在实际的京剧服饰文化翻译过程中，正如苏联翻译家巴尔胡达罗夫在《语言与翻译》一书中所表达的观点。译者首先应从文化翻译的战略角度去考量翻译内容，遵循文化翻译协调原则，坚持"恰恰调和"和"原汁原味"的文化翻译，并在基于理解和转换的基础上，对译文读者或观众获得的审美感受与认识做细化整理。因为从翻译开始到终结，从阿尔法到欧米茄，上下、纵横、左右、前后都要协调，各个层面都要协调，各种关系都要协调，无数的矛盾、无数的冲突都要求译者尽量去协调。翻译本身就是协调，是不同语言因素之间相互协调、相互让步、相互迁就和相互融合的实施与结果。协调贯穿于翻译的整个过程，自始至终。翻译的过程，包括翻译的每一阶段，都是协调的实施与结果。❶

翻译是一种语言转换，服饰艺术表达需要借助语言手段，其中包括带有浓厚文化底蕴的特色词汇。纵观我国翻译史，佛经的翻译、徐光启的科技翻译、严复的社会政治翻译都存在翻译的文化战略考虑。首先，就文化翻译审美原则而言，严复、林语堂、赵元任、胡适、傅雷、林以亮、钱钟书都可以说是先驱者。❷其次，在文化翻译原则方面，即要遵循西方文化翻译理论主张的异化翻译观，尽可能对文化个性给予保留。❸

（三）京剧服饰文化协调翻译实施策略

在京剧服饰文化艺术翻译过程中，转换并非易事，而且原作与译作之间的转换也并非是在理想环境或真空中进行的文本转换，而在具体的文化语境中，还有各种文化的"干扰"，使翻译自始至终都受到民族文化的影响，原作的选择和译语文化的需求也同样密不可分。因此译者翻译策略的确定既受到译语文化的影响，也与翻译所要达到的交际目的有关，译作的被接受是两种文化的碰撞后文化适应的结果。原作毕竟是另一民族的文化产品，如何将它转换、移植到译语文化中，并且避免文化冲突的产生，是每个译者在文化协调中亟待解决的关键。

如何协调？关键在于协调。从理论角度而言，协调是一种指导人们在翻译实践中如何求同减异、做好沟通的翻译理论。其目的在于帮助解决翻译的操作过程中两种语言间一切需待重新磨合与适应的问题，因为非此，则无法求得翻译上成功的转换。其必要性与前提是两种语言间一般等同性的存在与可译性（而翻译的要求则是其天然的求同性）；其可能性是翻译转换本身所允许的一定的通融性及其差异的容许量，因而"度"的适量把握势必会成为操作时的首要要求。其次，协调的运用范围之广阔，遍及译语的各个成分与层次的一切方面以及全部过程。在这个意义上，它属于翻译理论中

❶ 张慧琴. 高健翻译协调理论研究［D］. 上海：上海外国语大学，2009：17.

❷ 刘靖之. 翻译——文化的多维交融(《文化翻译论纲》代序)［J］. 中国翻译，2000 (1)：22–26.

❸ MUNDAY, J. Introducing Translation Studies：Theories and Applications［M］. New York：Routledge，2001：146–148.

实用性较强的重要层面。最后，在方法论上，它属于理论范畴，而非直接具体的翻译技巧，因而，该理论不具规定性而仅具指导性。❶翻译协调理论实质上是一种哲学的产物。在思想上，以认知论和方法论的双重性为基准，使理性与感性、分析与直觉、实证与思辨、推论与感悟、抽象与具体……所有这一切都贯穿于前述中这二者的相对并重与交互依存。但又在这每对矛盾中给予其后者以更为优先的着重，并不是只认其一而废弃忽略另一，更不走极端。在现象的对待上，则是颇具动感意识的，是带有开放性和兼顾性的，往往是从多方面考虑的；因而其方法也是多维的、多关系的、多因素的、多层面的和多向性的。这也正是翻译协调理论的科学性所在。

协调是一种本领、一种艺术，其关键在于审时度势、灵活把握。首先需要审读原文，基于民俗传统、宗教信仰、伦理道德和价值观念准确理解原文信息，综合各种服饰文化信息概括总结做出预断，充分考量不同文化中存在的差异，针对词义色彩、文化历史、问题章句全面分析，尝试对其表层的和深层的文化内涵给予辨析，实现"移情感受"，使主体孜孜于移客体之情于己、移客体之志于己、移客体之美于己，最终使达致物我同一，恰如《易经》中所谓"同声相应，同气相求"。对于翻译中遇到的问题，做到了然于心，包括哪些问题必须采取协调变通手段，是否具备协调转换的条件，怎么把握协调的时机和幅度，如何协调变通才能解决。

正如余光中先生所说，"翻译如同婚姻，是一门两相妥协的艺术"❷，要在众多的文化差异中协调，使译文至善至美，为读者理解接受，就必须理解文化差异、尊重文化差异、协调文化差异。在服饰文化艺术翻译过程中，转换并非易事，而且原作与译作之间的转换也并非是在理想的或真空中进行的文本转换，而是发生在具体的文化语境中，各种文化的"干扰"使翻译无时无刻都受到民族文化的影响，包括原作选择和译语需求而进行的调整。因此译者翻译策略的选择既要考虑译语文化的影响，也要考虑翻译所要达到的交际目的，译作的接受与否取决于两种文化的碰撞与适应。原作属于不同民族的文化产品，如何转换移植到译语文化中，尽量避免文化冲突，成为译者在文化协调中协调的关键。

吴志杰曾就中国传统文化中最能体现和合精神与和合价值的"意、诚、心、神、适"五个核心范畴❸，在《中国翻译》上分别聚焦翻译本体、翻译伦理、翻译认识论、翻译审美和翻译文化生态❹，给予系统分析与阐释，并认为上述五个范畴各自所对应的语言属性、伦理属性、思维属性、审美属性和文化属性，犹如五朵花瓣，使翻译之美蕴含其中。笔者结合服饰文化翻译特点，认为文化差异的协调处理是翻译协调的核心与关键。协调幅度的大小将直接关系到文化翻译的过程与结局，而服饰文化翻译过程

❶ 张慧琴. 高健翻译协调理论研究［D］. 上海：上海外国语大学，2009：18.
❷ 余光中. 余光中谈翻译［M］. 北京：中国对外翻译出版公司，2002：55.
❸ 吴志杰. 和合翻译研究刍议［J］. 中国翻译，2011(4)：6.
❹ 吴志杰. 和合翻译学论纲［J］. 广西大学学报(哲学社会科学版)，2012(1)：101.

中的"意、诚、度、心、神"思维层面的五步协调融合,阐释了文化协调的本体观、认知观和审美观;揭示了文化协调翻译的前提与条件、标准与要求、幅度与范围;凝聚着译者的苦心;浓缩了翻译的过程;阐释了译者尊重原文,以及在整个翻译过程中从原作到译作的曲折与妥协、磋商与纠结、协调与和谐。译者只有怀揣至"诚"之心、注重"度"的把握、讲究"神"的韵味,才能使译文在尊重原文的基础上实现"心"照不宣的"意"义转换,"意、诚、度、心、神"五者间的关系如图6-1所示。笔者对其逐一剖析如下:

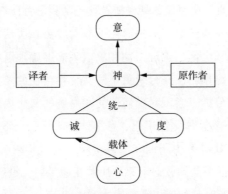

图6-1 服饰文化翻译中思维层面协调融合

"意"是译者对于原作本意的理解与揣摩,是情真意切的聚焦,是传情达意的先驱。从创作实践来说,"意"是指作家作为创作主体之"意识""意念""意向""意图""意志",乃至"心意""情意"等。它不一定是明确的理性意识,可以是一种潜在意识,是理性意识与感性认识的交融,是一种认知与情感的心意状态。❶我国古代美学思想中所谓"意在笔先",就是强调内容对形式的决定作用,"意"是提炼、选择艺术形式的依据,翻译过程中要"意在笔先,意在笔中,意在笔外",将"意"融化到笔墨中,做到"心手相师",以意感人。中国传统美学中,"意"是艺术美的精髓。❷在南朝宋时著名史学家范晔的《狱中与诸甥侄书》和唐朝杰出诗人、散文家杜牧的《答庄允书》中,都有关于"意"的表述,"以意为主"形成主轴,以言表意述志。"意"包含三个所指和三层意思:一是指内容,所谓"意定形随"(内容决定形式)或"形随意定"(形式无法超越内容);二是指情意;三是指意念,就是指在交流中的人要用自己的意念、意识去意会、解读对方之志。❸不同民族文化的人领悟到的意象、意境和意蕴可谓"不约而同",不同语言文化的人都会有其各自的"其味无穷",但是无论如何,"意"必须基于原文。

"诚"是关键,是核心,是灵魂。翻译的实质是意义的对应转换、是修辞立诚,《周

---

❶ 王秉钦. 翻译:模糊艺术 [J]. 上海翻译. 2009(11):21.

❷ 杨辛,甘霖. 美学原理新编 [M]. 北京:北京大学出版社,1996:130.

❸ 刘宓庆. 翻译美学导论(修订版)[M]. 北京:中国对外翻译出版公司,2005:iii.

易·乾卦》云："君子进德修业。忠信，所以进德也；修辞立其诚，所以居业也。"❶这就要求译者在翻译中超越语言层面的文字转换活动，把翻译与修德、立业、做人联系起来，这样才能真正处理好翻译中的伦理关系，才能解决好翻译中所遇到的种种矛盾与困难。我国历史上儒家的"温柔敦厚"、道家的"质朴深邃"、法家的"至诚之理"和墨家的"情理并茂"，都是"诚"的典范。"道"的真意包含了"诚"，意即理性行动，以开放的心态待人接物，尊重文化差异和异质他者。"诚"是一切翻译活动得以进行的前提，也是确保翻译顺利进行的保障。"诚于译事"既是一种做人的态度，也是从事翻译活动的前提和要求。"诚"也是一种开放、包容、积极、审慎的翻译态度与观念，开阔的胸襟，容纳原文作者和译者的表达习惯与文化差异，以至"诚"之心使主体达到对事物及他人的认识，以促进"诚"的进步发展，从而形成译文与原文之间的和谐沟通。翻译脱离不开语言文化发展的基本轨迹，翻译不能不顾源语，也不能不关照目的语。译者只有以坦诚的胸襟，尊重原作，诚心诚意传递原作的本意，才能真诚负责地面对原文的作者和译文的读者。离开原作原意的艺术加工，一味追求"华藻"，不顾原文语义的限制，也不论译者的用意何在，都是对不谙原作的读者的不负责任与误导。因此，"诚"也是译者自身道德水准与知识水平不断提升、真诚对待翻译中诸多复杂的关系和矛盾的修炼过程。大师级的翻译家绝不仅仅是一个匠人，必定是一个有天赋又勤奋的德艺双馨的译者。❶

"度"是翻译活动的原则和策略，更是翻译活动的前提。因为翻译活动本身也是人类生存的一种方式。人类以生存为目的，其第一个自觉意识是"度"的意识。科学、艺术、文化和哲学其实质都是人类"度"的意识的实现与客观化。翻译活动直接涉及两种事物的对比和转化，对"度"的要求程度越高，其艺术性表现就越为突出。"度"和实践紧密联系，翻译实践要求把握分寸恰到好处，注重对"度"的追求和把握。译者只有把握分寸，结合原作和译文读者所处的社会背景，充分考虑文化差异，掌握"度"的范围和标准，才能灵活有度、协调适度。而译作对于原作需要给予"适量、适时、适度"的协调，"适量"是指协调的范围不允许太广、不允许超量，否则译文与原文之间势必导致差异巨大，协调过量有改写之嫌，协调不足，放不开手脚，又可能导致译文读者一头雾水，不容易明白；"适时"是指协调的时间与场合、社会环境与时代要求、读者需求与社会需要等之间都要"恰到好处"，可谓做到真正意义上的"诚"、名副其实的"忠"。

"心"是"诚"与"度"的载体，是我国文化思想史上极其重要的哲学概念，是哲学、美学领域不可或缺的文化积淀。"心"的文化意义经历了由生理而文化的历史延脉。《说文解字》对"心"的解释为："心、人心、土藏，在身之中。象形。博士说以为火藏。"这里的"心"是一种生理意义上的存在。《黄帝内经》中说："夫心者，五藏

---

❶ 吴志杰，王育平. 论"合和"本体的非实体性特征［J］. 湖南科技大学学报，2009a(6)：140.

之专精也。"中国古代曾长期把思维的能力归于心之功能，直到明代李时珍提出"脑为元神之府"（《本草纲目·木部》），清代王清任提出"灵机记性不在心在脑"（《医林改错》）的"脑髓说"，传统的认识才逐渐得到纠正。"心"由思维的器官也就逐渐演化成了哲学、美学的范畴。荀子认为，心生而有知，知而有异，异也者，同时兼知之。心求道，心道合一，才能达到至人的境界。❶在翻译中，只有用心、专心和诚心，心源为炉，造化神秀，才能美善合一，美善合一；对原作者怀有敬畏之心，才能静心、安心地深度挖掘、认识、理解服饰文化内涵，将心比心，换位思考，做到与原文作者心照不宣、心心相印，成就默契和谐的译作，尽心尽力用译语再现原语之美。

"神"是"心"之精所发出的活动，"心、诚、度"协调统一的结局，是原文作者和译者之间坦诚相待、心照不宣，心有灵犀、形神融合的结晶，是原作意义的升华，是原作新生的体现，是译者对于译作风格的审美把握，是译作对于原作灵魂的容光焕发与神采飞扬。服饰文化艺术中所体现出的"神"，实际是译者所了解和经过情感体验的"神"，"传神"的译作体现了原作的本质特征。

服饰文化艺术翻译中"意、诚、度、心、神"五层面的协调融合，都是基于对原文整体的理解与把握，犹如乐曲的创作与演奏，设定意向，描摹意象，调节音调、音量与音质，注重全局性的艺术手段，不断"微调"，聚焦原文文化信息的特征，针对翻译中的失色（艺术着色的力度不够与"力度过剩"），采取淡化与强化的手段，在反复经营中不断协调，择善从优，最终实现笔底生花，使目的语读者感受到异域文化带来的愉悦。

（四）京剧服饰文化翻译"真、善、美"的协调统一

服饰作为特殊的文化载体，服饰文化翻译是一项特殊的文化艺术翻译实践，是"真、善、美"和谐统一的体现。法国古典主义美学家布瓦罗（Boilean）强调"真"与"美"的关系，认为两者和谐统一。❷古希腊哲学家毕达哥拉斯（Pythagoras）认为，"美"就在内外和谐的契合。先秦哲学家庄子强调"法天贵真""原天地之美而达万物之理"。古希腊哲学家苏格拉底（Socrates）认为"美在善"，18世纪法国唯物主义美学家狄德罗（Diderot）认为："'真、善、美'是紧密联系在一起的，在'真'或'善'上加上某种罕见或令人注目的情景，'真'就变成'美'了，'善'也变成'美'了。"❸甚至历史上法国的浪漫主义的三大倾向，包括努力忠实地再现过去历史的某一片断或现代生活的某一侧面——"真"的倾向；努力探索形式的完美，把它领悟为表现方面的仪态万千、历历如画，或者音律方面的严格及和谐，或者一种由于简洁单纯而不朽的散文

---

❶ 刘桂荣. 徐复观美学思想研究［M］. 北京：人民出版社，2007：84.

❷ 刘宓庆. 文化翻译论纲［M］. 武汉：湖北教育出版社. 2005：247.

❸ 叶立诚. 服饰美学［M］. 北京：中国纺织出版社，2007：62.

风格——"美"的倾向；热衷于伟大的宗教革新观念或社会革新观念，即艺术中的伦理目的——"善"的观念。❶ 这三种主要倾向中的每一种倾向都产生了价值伟大而持久的作品，都从不同角度体现了"真善美"的协调统一，具体的服饰文化翻译协调全过程如图6-2所示。

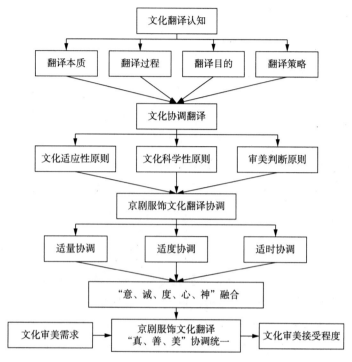

图6-2 京剧服饰文化翻译协调全过程框架

京剧服饰文化基于对文化翻译的认知，探究翻译的本质、翻译的过程、翻译的目的与策略，尝试实施文化协调翻译，并始终坚持遵循文化适应性原则、文化科学性原则与文化审美判断原则，在协调过程中注重把握协调内容量的多少、协调幅度范围的大与小，根据协调的时间、场合与环境适时协调，在思维层面进行"意、诚、度、心、神"的五步融合，充分考虑译文读者的文化审美需求与文化审美接受程度，协调多对矛盾，在服饰文化翻译的全过程中达到"真、善、美"的协调统一。

服饰文化作为文化艺术的组成，京剧服饰文化艺术的翻译实践本身属于不同文化之间的对话，这就意味着首先做到不要以本位文化作为文化的起始点和归宿，而应该最大程度努力忠实地再现过去历史的某一片断的服饰文化，或现代生活的某一侧面的服饰文化美。

---

❶ 许渊冲. 翻译的艺术［M］. 北京：五洲传播出版社，2006：285.

京剧服饰文化翻译理论探索 第六章

213

（五）京剧服饰文化协调翻译的标准

翻译是文化的产物，是人类社会一项极其古老的文明活动，它几乎与人类文明史同时诞生。原始部落之间的彼此往来、互通有无以及各个时期典籍的传播机制，包括当今世界各国各民族之间科学、艺术、历史、哲学、科技、政治和经济的频繁交流都仰仗翻译来实现。翻译不仅仅是人类生存的果仁和甘泉，还是人类思想和精神的阳光与空气。翻译活动使各个民族摆脱狭隘的地域限制，开拓了人们的视野，增进了不同文化和价值观念的借鉴与交流。翻译带来的不同文明成果的交汇有利于世界各民族之间的交往与和平相处，同时促进了人类文明的共同进步和繁荣。在实际的翻译过程中，借助融合或互补的翻译方法，通过译文与原文之间的相互让步、妥协、迁就、牺牲、凑合、两全、折中，甚至是"和稀泥"，达成了原作与译文读者之间的彼此和谐共处；❶在尊重差异、正视差异的同时，心平气和地协调处理文化差异，去粗取精，去伪存真，由表及里，尽量对原文的信息进行"翻译审美艺术表现手法"，使服饰文化翻译艺术化。同时，更要坚持文化协调观，针对依照译文与原文之间文化差异的大小，采取不同幅度的协调，既尊重文化的民族性，又体现文化的兼容性，在文化"保真"中自觉与不自觉地实现彼此文化的创新、丰富与发展。

京剧服饰作为特殊的文化载体，不同文化之间的对话不能以本位文化作为文化的起始点和归宿（求真），而是要以平等的态度、开放的心态互相学习（求善），取长补短，提高对"他者"的敏感度。了解文化"他者"，无疑有助于"自我"反省和"自我"完善，以便加强"自我"意识，更好地"自我"定位。研究"他者"，最终目的在于更好地理解发展"自我"（求美）❷。在服饰文化翻译中，要深入探究原文的本意，虚心求教，最大程度求真；同时，满怀善意，换位思考，充分考虑读者的需求、接受能力和社会发展的主旋律，要为社会的进步输入正能量。只有这样才有可能成就忠实于原文、源语文化和译语文化之间的差异协调，体现中华传统服饰文化内涵协调美的理想译文。协调是手段、是方法、是原则，而和谐的译文以及能促进不同民族服饰文化交流发展的译文才是文化协调翻译的不懈追求与最高标准。

本章小结 ｜ 在京剧服饰文化翻译中，只有深入探究原文的本意，虚心求教，最大程度求真；同时，满怀善意，换位思考，充分考虑读者的需求、接受能力和对外传播的主旋律；以及顺应翻译生态环境，恪守"真、善、美"的协调统一，正确选择符合现代人类审美需求的中华服饰文化元素，为人类的和平与发展输入正能量，才有可能实现忠实于原文，协调源语文化和译语文化之间的"文化差"和"语言差"。不去追求大而全的"原汁原味"，而是选择不同程度的协调，以部分代整体，通过不间断的

---

❶ 张慧琴. 高健翻译协调理论研究 ［D］. 上海：上海外国语大学，2009：28.

❷ 张慧琴.《论语》"以'礼'服人"服饰文化英译探索 ［J］. 中国外语，2014，11(06)：84.

努力，从点滴做起，把服饰文化介绍给世界，才能最终成就中华传统服饰文化内涵美的理想译文。同时，文化本身具有自我丰富的功能，随着世界文化的进一步融合，不同文化之间对于原语文化空白的理解也会随着时间的推移发生变化，文化翻译的内容也应不断调整。这就要求译文应该提高对文化的敏感性和自觉性，不断丰富自己的文化知识才可以肩负起跨文化交流的重任。在文化语境的翻译过程中，也要清醒地认识到，京剧服饰文化的翻译标准与结果也不是一成不变的，需要结合历史和时代背景以及地域文化特征、风俗习惯等，以动态的视野与开放的心态不断顺应历史与社会的需要，协调各种差异；以尊重、理解、接受的平和心态换位思考；用宽广的胸怀与国际化的视野看待差异、理解差异、接受差异、协调差异；以一名文化人的身份审时度势，既要实现保留、传播、发扬本民族优秀的文化遗产和对历史负责，也要不断吸纳不同国家和民族的优秀文化，对时代负责，传播京剧服饰文化，助力传统京剧的对外传播，增强国人文化自信，不断丰富、发展自身的传统文化，使京剧服饰文化翻译成为译者探索、实践与传播京剧服饰文化艺术"真、善、美"的幸福历程。

第七章

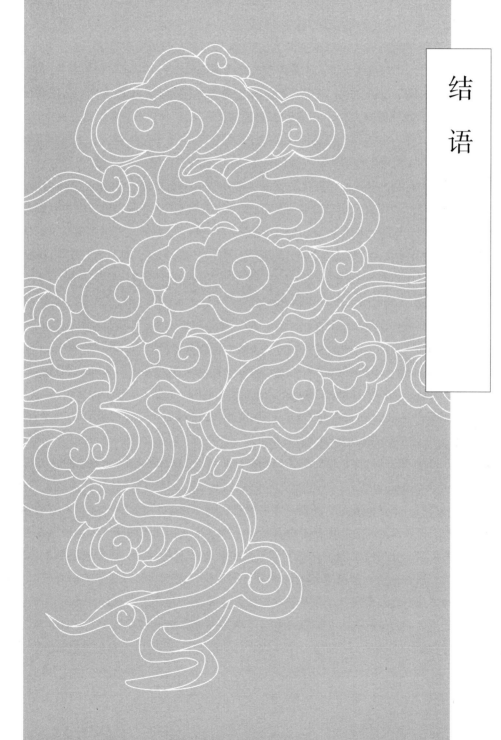

结 语

京剧艺术具有独特的审美价值，其特定的程式不仅表现在唱、念、做、打之中，也表现在服饰与化妆之中，并成为京剧舞台塑造人物形象必不可少的手段。

本书博取众家之长，在基于京剧服饰分类的基础上，依照大衣、二衣和三衣各自所包含的戏剧服饰文化，汇集梳理，采用汉语概述、中外学者阐释、英汉文本对照，并基于文化内涵的深入剖析，注重传统服饰文化的传承与传播，从不同角度和多层面再现京剧衣箱的发展演变、应用及其功效。

无独有偶，对于戏曲服饰的重视，中外皆然，而文化的传播也恰恰得益于精美服饰带给观众的深刻记忆。著名的莎士比亚戏剧，对于戏曲服饰的重视与我国传统京剧相比，则是有过之而无不及。以《罗密欧与朱丽叶》为例，两位主角从出场的服饰就传递出了性别、家境、社会地位等，下面不妨同样采用英汉对照的方式予以呈现。

原文 To begin, we know that Romeo was dressed as a man and Juliet as a woman. Certainly this is stating the obvious, however since all of the actors on the Shakespearean stage were male. Their characters, sex or gender resulted overwhelmingly from the clothes they wore. Juliet's feminine beauty differed from Romeo's masculine attractiveness only because the costumes they wore made it so. Next, their clothes presented Romeo and Juliet's social station as the eldest children of wealthy, prominent families. Early modern England was a highly stratified society that was well versed in the apparel that marked an individual's place in the hierarchy. Audiences would have noted the elite apparel in which Romeo and Juliet appeared and recognized that each was the other's social equal.❶

拙译 剧目一开始，我们从不同的扮相就知道了罗密欧和朱丽叶各自的性别。虽然莎士比亚戏剧舞台上的所有演员都是男性。但戏曲角色的性格、性别主要是通过服饰来体现，朱丽叶的女性柔美不同于罗密欧的阳刚魅力，这些都借助服饰给予传递。同时，戏曲角色人物的服装也展现其社会地位，朱丽叶和罗密欧都出身于名门望族，在家里分别是长女和长子。早期英国的社会等级分化严重，服饰便成为不同人群在等级制度中地位的标识。观众通过罗密欧和朱丽叶出场时的着装，便可知道两人的社会地位平等。

❶ LUBLIN R I. Costuming the Shakespearean Stage：Visual Codes of Representation in Early Modern Theater and Culture ［M］. Farnham：Ashgate，2011：3.

原文 Clothing and theatrical apparel carried specific meanings that were well understood by the playwrights, performers, and audiences of the time. As Stephen Greenblatt explains, "what can be said, thought, felt, in this culture seems deeply dependent on the clothes one wears—clothes that one is, in effect, permitted or compelled to wear, since there is little freedom in dress."❶

拙译 服装和戏剧服装都具有人们熟知的特定含义，为当时的剧作家、演员和观众所认同。正如斯蒂芬·格林布拉特（Stephen Greenblatt）解释的那样，"在戏剧文化中，角色所说、所想和感受，似乎都取决于其所穿服饰。实际上，这些服饰的穿用完全服从于戏曲要求，而演员本人对于服饰没有任何选择权。"

原文 To the early modern English audiences that attended the theater, costumes engaged the issue of social station on a number of levels. Beyond simply marking a character's place in the hierarchy of a particular play, the clothes worn in production also actively participated in establishing expectations, denoting power relations, delimiting freedom, prompting action, and more.❷

拙译 对于早期观看戏剧的现代英国观众来说，服装在诸多层面暗示了角色的社会地位。除了定位角色等级身份之外，它还积极表达了人们的期望、暗示了权力、圈定了自由以及行动的前奏等。

综上所述：京剧服饰作为特殊的文化载体，其文化翻译是一项特殊的文化艺术翻译实践。全书在写作过程中，笔者采用英汉对照的方式，聚焦大衣、二衣和三衣及其配饰，归类挖掘与剖析京剧服饰文化内涵及其文化翻译的英文表达，借助他译与拙译论证并践行"真、善、美"服饰文化协调翻译理论，通过采用灵活多样化的协调方式，如删译、截译、增译、改译、补译，乃至拼音翻译等，协调译文与原文的差异。对于具有中国特色的正能量的文化词汇，尽量采用汉语拼音音译、音意兼译或定语修饰的译介方法，❸忠实于原文"求真"；同时换位思考，体谅当今文化语境下西方读者对于京剧服饰文化的审美需求，包括接受能力与兴趣，尊重原作，关注观众和译文读者的需求，怀揣至诚之"善意"；充分考虑京剧服饰带给观众的无限遐思与美妙联想，特别是对于展示人物性格、推动故事情节和深化故事主题等诸多方面的总和效用。在译文的表达上，既要准确，还要不啰唆；既要注重中华服饰文化的传承，还要详略得当，换位思考。如此的前后思量，既是对传统服饰文化的挖掘与珍视，也是中外文化交流的

❶ STEPHEN GREEN BLATT. "General Introduction" in the Northern Shakespeare. 2nd ed. New York：WW Norton，2008：59.

❷ LUBLIN R I. Costuming the Shakespearean Stage：Visual Codes of Representation in Early Modern Theater and Culture［M］. Farnham：Ashgate，2011：78.

❸ 张彩霞，朱安博. 负面意义的汉语借词及话语权［J］. 上海翻译，2016(3)：55.

体验与实践，更是传统京剧服饰文化中"真、善、美"和谐统一的再现。

同时，由于京剧服饰文化博大精深，吸引中外学者从不同角度对其研究。本书列举了部分国外京剧爱好者对于服饰文化的介绍，并通过笔者的"拙译"，可以看到国外学者对于传统服饰文化内涵的解读程度，虽然有些地方不够全面，但是没有任何偏颇之处，可谓当时历史时期的"完美"英文表达，这对于京剧爱好者学习理解京剧服饰，尝试从异域文化的角度掌握其英文表达具有一定的帮助。书中对于部分学者针对同一类服饰英文表述的不同对比，以及补充性的建议或评价，旨在从不同角度理解、丰富服饰文化英文表达。尤其是本书中引用了很多相关专业书籍的服饰介绍文本，包括英汉平行文本，并针对一些具体表述给以补充或阐释，大多采取总体简要评述，而非逐一的评述与剖析，更希望留给读者自己评判赏析。

京剧服饰文化翻译任重道远，理论探索与构建更是各持己见。翻译协调理论虽然已经在翻译家高健先生毕生的翻译实践中论证，也得到学界诸多专家的肯定，但是在京剧服饰文化中指导意义与价值依然需要在时间的长河与岁月的磨砺中，不断论证检验。我们有理由相信，和谐的理念、协调的手段与方法，终究会成就服饰文化和谐美的译文。京剧服饰也必将在文化翻译协调理念的不断探索中，在构建人类命运共同体的和谐目标中，以其独特的和谐美，大放异彩。

参考文献

[1] ANDRE LEFEVERE. Translating Literature: Practice and Theory in a Comparative Literature Context [M]. Beijing: Foreign Language Teaching and Research Press, 2006.

[2] BANU G, WISWELL E L, GIBSON J V. Mei Lanfang: "A case against and a model for the occidental stage" [J]. Asian Theatre Journal, 1986, 3(2): 153-178.

[3] BRANDON JAMES R. The Cambridge Guide to Asian Theatre [M]. Cambridge: Cambridge University Press, 1993.

[4] BUSS K. Studies in the Chinese Drama [M]. Boston: The Four Seas Press, 1992.

[5] DOLBT W. Father of Chinese Drama: A Sketch of the Life and Works of Guan Hanqing [M]. Edinburgh: U of Edinburgh Press, 1983.

[6] DOLBT W. Eight Chinese Plays from the Thirteenth Century to the Present [M]. New York: Columbia University Press, 1978.

[7] DOLBT W. West Wing/by Wang Shi-fu [M]. Edinburgh: Caledonian Publishing Company, 2002.

[8] FENG MENGLONG. The Perfect Lady by Mistake and Other Stories [M]. London: P. Elek, 1976.

[9] GARRETT V M. Traditional Chinese Clothing: in Hong Kong and South China, 1840-1980 [M]. Hong Kong: Oxford University Press, 1987.

[10] GEORGE, SOULIE DE MORANT. Théâtre et Musique Modernes en Chine [M]. Paris: P. Geuthner, 1926.

[11] HSU YI-LIN. A Comparison of the Vocal Techniques in Peking Opera and Bel Canto Opera [D]. Santa Barbara: University of California, Santa Barbara, 1992.

[12] HWANG MEI-SHU. Peking Opera: a Study on the Art of Translating the Scripts with Special Reference to Structure and Conventions [D]. Tallahassee: The Florida State university, 1976.

[13] LEETONG SOON. Chinese Theatre, Confucianism, and Nationalism: Amateur Chinese Opera Tradition in Singapore [J]. Asian Theatre Journal, 2007, 24(2): 397-421.

[14] MACKERRAS C P. The Chinese Theatre in Modern Times: from 1840 to the Present Day [M]. London: Thames and Hudson, 1975.

[15] MACKERRAS C P. The Rise of the Peking Opera, 1770–1870: Social Aspects of the Theatre in Manchu China [M]. Oxford: Clarendon Press, 1972.

[16] MACKERRAS C P. Chinese Drama: a Historical Survey [M]. Beijing: New World Press, 1990.

[17] MACKERRAS C P. Amateur Theatre in China 1949–1966 [M]. Canberra: Australian National University Press, 1973.

[18] MACKERRAS C P. Theatre and the Taipings [J]. Modern China, 1976, 2(4): 473–501.

[19] MACKERRAS C P. Chinese and Western Drama Traditions: a Comparative Perspective [C] // China and Other Spaces: Selected Essays by Contributors to the Research Seminar Series of the Institute of Comparative Cultural Studies at the University of Nottingham Ningbo, China 2005–2007 [M]. Nottingham: Critical, Cultural and Communications Press, 2009: 21–34.

[20] MACKERRAS C P. Yangzhou Local Theatre in the Second Half of the Qing [C] // Lucie B. Olivová, Vibeke Børdahl(Eds). Lifestyle and Entertainment in Yangzhou [M]. Copenhagen: NIAS Press, 2009.

[21] MACKERRAS C P. Images of Asia: Peking Opera [M]. Hong Kong: Oxford University Press, 1997.

[22] MAILEY J. Embroidery of Imperial China [M]. New York: China Institute in America, 1978.

[23] NICOLAS BOILEAU. L'Art Poétique de Boileau-Despréaux ed. G. H. F. de Castres [D]. Leipzig: Librairie de E. Wengler, 1856.

[24] SCOTT A C. The classical Theatre of China [M]. London: Allen & Unwin, 1957.

[25] SCOTT A C. An Introduction to the Chinese Theater [M]. Singapore: D. Moore, 1958.

[26] SCOTT A C. Chinese in Transition [M]. Singapore: D. Moore, 1958.

[27] SCOTT A C. Mei Lanfang, Leader of the Pear Garden [M]. Hong Kong: Hong Kong University Press, 1959.

[28] WILLAMS C A S. Chinese Symbolism and Art Motifs: a Comprehensive Handbook on Symbolism in Chinese Art through the Ages [M]. 4th ed. North Clarendon, Vt.: Tuttle Pub, 2006.

[29] WICHMANN E. Traditional and Innovation in Contemporary Beijing Opera Performance [J]. TDR(1988–), 1990, 34(1): 146–178.

［30］WICHMANN E. Listening to Theatre：the Aural Dimension of Beijing Opera［M］. Honolulu：University of Hawaii Press，1991：1-525.

［31］WICHMANN E. They Sing Theater：the Aural Performance of Beijing Opera［D］. Honolulu：University of Hawaii，1983：1-525.

［32］WICHMANN E. 当代京剧演出中的传统与革新［J］. 马海玲，译. 戏曲研究，1993 (44)：202-220.

［33］WICHMANN E. 在国外演出京剧的翻译与导演工作［C］. 北京：中国艺术研究院，1987.

［34］WU ZUGUANG，HUANG ZUOLIN，MEI SHAOWU. Peking Opera and Mei Lanfang：A Guide to China's Traditional Theatre and the Art of Its Great Master［M］. Beijing：New World Press，1981.

［35］ZUCKER A E. The Chinese Theater［M］. Boston：Little，Brown，and Company，1925.

［36］卞赵如兰. 西方对于京剧研究的情况［C］//中国戏曲艺术国际学术研讨会论文集. 北京：中国艺术研究，1987：72-87.

［37］曹广涛. 英语世界的中国传统戏剧研究与翻译［M］. 广州：广东高等教育出版社，2011.

［38］何九盈，王宁，董琨，等. 辞源［M］. 3版. 北京：商务印书馆，2015.

［39］赫伯特·A. 西蒙. 管理行为［M］. 杨砾，韩春立，徐立，译. 北京：北京经济学院出版社，1988：44.

［40］胡庚申. 从术语看译论——翻译适应选择概论［J］. 上海翻译，2008(2)：1-5.

［41］康保成. 中国近代戏剧形式论［M］. 桂林：漓江出版社，1991.

［42］张慧琴；徐珺. 文化语境视角下的《红楼梦》服饰文化汉英翻译探索［J］. 山东外语教学，2013(5)：98-101.

［43］张传官. 急救篇校理［M］. 北京：中华书局，2017.

［44］张晓娜. CNNIC发布第40次《中国互联网络发展状况统计报告》［N］. 民主与法制时报，2017-08-08.

后 记

时光荏苒，开笔写这本书的时间应该是2015年的冬天，虽然在这个过程中也完成了一些其他重要的书稿，甚至也还在继续着更为艰难的论著，但这本书的完成中着实有着太多让我刻骨铭心的记忆。

首先是教学相长，该课题的研究过程也是师生共同成长的历程。研究生孙鸿翔、赵苗都对京剧服饰术语的协调翻译进行了资料的收集、整理与汇编，研究生张家琦在通读全书的过程中，对全书的"参考文献"给予了全面调整，并结合译文中的部分节译，甚至包括个别原文中的拼写问题，先后请教了北京京剧院主任舞台技师、剧装总监彭仲林先生，北京市非物质文化遗产项目"京剧盔头"的第五代传人、出生于盔头制作世家的李鑫老师，以及京剧脸谱绘制世家的传人赵楠老师。同时，在该书的撰写过程中，我们还请教了精通京绣、珠绣、苗绣和戏剧服装制作等非遗技艺的传播者安宁老师和毕业于中国戏曲学院的程桂奇。这些专家的意见或建议斧正了一些京剧服饰表达在一定程度上"流传已久"的瑕疵或"以讹传讹"的"笔误"。正是大家的合力与项目基金的资助才促成了本书的顺利出版。

其次是有关京剧服饰文化翻译的理论探索与建构。翻译理论在一定程度上类似于数学中的公理或定理，对于初出茅庐、名不见经传的"晚辈"怎敢"狂言"？但是翻译协调理论毕竟是笔者二十余年字斟句酌、反躬自省、坚持不懈探索的主题，也是教学与科研融合历程的见证，更是对于恩师——翻译家高健先生，其毕生翻译实践中逐步形成的翻译思想的梳理与提升。或许有学者会质疑该理论的适用范围和程度，甚至会说该理论源于高健先生散文翻译实践，怎么证明能适用于京剧服饰文化翻译？事实上，学界对于理论的探究历来众说纷纭，好在该书倡导的文化翻译协调理论是在诸多学者翻译实践基础上的反思，正如许多大家所言，翻译就是穷对付，翻译就是折中的艺术，翻译就是协调。或许也有学者认为尝试构建的翻译理论应该是界定在汉语翻译为英语，顺应文化"走出去"。但是笔者认为，我们今天的"走出去"正是基于原有的"已出去"，对照反思"已出去"才能看清哪些文化内涵有待于我们挖掘，才能更好地

"走出去"。这个过程复杂艰辛，在一定程度上本书也只是选取了蟒和盔帽两类服饰的不同英译文本进行较为详细的比较研究与探究文化词汇的选择，特别是针对"已出去"的文化内涵表达，笔者通过"拙译"和"补译"的方式方便读者对照反思，揣摩提炼，从中英不同角度更为全面地理解京剧服饰文化内涵。但是相对于全书大衣箱、二衣箱和三衣类的英汉对照文本，逐一剖析和比较则较为鲜见，姑且算是"留白"，供读者自行评判取舍；包括文化翻译协调理论在诸多京剧服饰双语对照表达过程中的具体运用，也并未给予阐释论证，其原因在于理论源于翻译实践的反思，而哲学层面的协调理论属于理念的指导，"意、诚、度、心、神"的步骤也属于思维层面的导引，翻译协调理论本身不具规定性，而仅具有指导性。在实际的翻译过程中，有学者可能会认为无非是"异化与归化"，这种说法在一定程度上完全正确，因为协调翻译是对于其他翻译理论的补充，本身并不排斥其他理论的存在与运用。但是在"异化与归化"这两大处理文化差异的法宝中，我们更倡导"优化"，而这"优化"则恰恰是异化与归化的并用，或者说是协调使用异化与归化两种翻译策略的结果与结局，是贯穿翻译过程始终的"意、诚、度、心、神"的协调产物。

最后，我想对亲爱的读者说，忙碌之余，让我们不妨多看看京剧，尝试从不同角度观赏、思考京剧服饰，在姹紫嫣红中品味平实恬淡的人生。谨以此书献给我过世已一年的慈父，天若有灵，祈愿治学严谨、一生勤勉的老人家能够感应到后辈们在求知与探索中的传承与努力。

张慧琴

2019年12月6日

于北京樱花园甲2号